New Technologies in Animal Breeding

Edited by

BENJAMIN G. BRACKETT
Department of Clinical Studies–New Bolton Center
School of Veterinary Medicine, and
Department of Obstetrics and Gynecology
School of Medicine
University of Pennsylvania
Kennett Square, Pennsylvania

GEORGE E. SEIDEL, JR.
Animal Reproduction Laboratory
Colorado State University
Fort Collins, Colorado

SARAH M. SEIDEL
Animal Reproduction Laboratory
Colorado State University
Fort Collins, Colorado

1981

ACADEMIC PRESS
A Subsidiary of Harcourt Brace Jovanovich, Publishers

NEW YORK   LONDON
PARIS   SAN DIEGO   SAN FRANCISCO   SÃO PAULO   SYDNEY   TOKYO   TORONTO

This book was produced in part from work funded by the Office
of Technology Assessment (OTA) of the United States Congress.
The views expressed do not necessarily represent those of OTA.

ACADEMIC PRESS, INC.
Orlando, Florida 32887

United Kingdom Edition published by
ACADEMIC PRESS, INC. (LONDON) LTD.
24/28 Oval Road, London NW1 7DX

Library of Congress Cataloging in Publication Data
Main entry under title:

New technologies in animal breeding.

    Includes bibliographies and index.
    1. Livestock--Breeding. 2. Breeding. I. Brackett,
Benjamin G. II. Seidel, George E., Jr. III. Seidel,
Sarah M. IV. Title: Animal breeding.
SF105.N484   636.082   81-20556
ISBN 0-12-123450-9   AACR2

PRINTED IN THE UNITED STATES OF AMERICA

    83 84   9 8 7 6 5 4 3 2

Contents

Contents

7. Preservation of Ova and Embryos by Freezing

S. P. Leibo

8. Applications of *in Vitro* Fertilization

Benjamin G. Brackett

9. Amphibian Nuclear Transplantation: State of the Art

Robert Gilmore McKinnell

10. Parthenogenesis, Identical Twins, and Cloning in Mammals

Clement L. Markert and George E. Seidel, Jr.

11. Gene Transfer in Mammalian Cells

Davor Solter

IV Analysis of Impact of Technology on Animal Breeding

12. Potential Genetic Impact of Artificial Insemination, Sex Selection, Embryo Transfer, Cloning, and Selfing in Dairy Cattle

L. Dale Van Vleck

13. Economic Benefits of Reproductive Management, Synchronization of Estrus, and Artificial Insemination in Beef Cattle and Sheep

E. Keith Inskeep and J. B. Peters

V   Continuation and Implementation of Research

14.   Continuation and Implementation of Research

*Benjamin G. Brackett, George E. Seidel, Jr., and Sarah M.
Seidel*

Contributors

Numbers in parentheses indicate the pages on which the authors' contributions begin.

*Keith J. Betteridge** (109), Agriculture Canada, Animal Diseases Research Institute, Nepean, Ontario K2H 8P9, Canada

Benjamin G. Brackett (3, 141, 257), Department of Clinical Studies–New Bolton Center, School of Veterinary Medicine, and Department of Obstetrics and Gynecology, School of Medicine, University of Pennsylvania, Kennett Square, Pennsylvania 19348

Wallis H. Clark, Jr. (91), Animal Science Department, University of California, Davis, California 95616

Robert H. Foote (13), Department of Animal Science, Cornell University, Ithaca, New York 14853

W. C. D. Hare (109), Agriculture Canada, Animal Diseases Research Institute-Nepean, Ontario K2H 89P, Canada

E. Keith Inskeep (243), Division of Animal Science, West Virginia University, Morgantown, West Virginia 26506

S. P. Leibo† (127), Biology Division, Oak Ridge National Laboratory, Oak Ridge, Tennessee

Ann B. McGuire (91), Animal Science Department, University of California, Davis, California 95616

Robert Gilmore McKinnell (163), Department of Genetics and Cell Biology, University of Minnesota, St. Paul, Minnesota 55108

*Present address: Centre de Recherche en Reproduction Animale, Faculté de Médecine Vétérinaire, Université de Montréal, C. P. 5000, St-Hyacinthe, Quebec J2S 7C6, Canada.
†Present address: Rio Vista International, Box 242, Route 9, San Antonio, Texas 78227.

Clement L. Markert (181), Department of Biology, Yale University, New Haven, Connecticut 06511

J. B. Peters* (243), Division of Animal Science, West Virginia University, Morgantown, West Virginia 26506

George E. Seidel, Jr. (3, 41, 181, 257), Animal Reproduction Laboratory, Colorado State University, Fort Collins, Colorado 80523

Sarah M. Seidel (41, 257), Animal Reproduction Laboratory, Colorado State University, Fort Collins, Colorado 80523

Thomas J. Sexton (81), Avian Physiology Laboratory, U. S. Department of Agriculture, SEA-AR, Beltsville, Maryland 20705

Elizabeth L. Singh (109), Agriculture Canada, Animal Diseases Research Institute, Nepean, Ontario, Canada K2H 89P

Davor Solter (201), The Wistar Institute of Anatomy and Biology, Philadelphia, Pennsylvania 19104

L. Dale Van Vleck (221), Department of Animal Science, Cornell University, Ithaca, New York 14853

*Deceased.

Preface

The desirability of reviewing new reproductive technologies in animal breeding became a necessity upon initiation by the Office of Technology Assessment of the United States Congress (OTA) of a study of "Impacts of Applied Genetics: Animal Breeding" to enlighten legislative decision-makers. In the course of this assessment, 37 experts representing diverse areas convened in Denver, Colorado, January 15–17, 1980, for a conference from which this book has evolved.

This volume is intended not to provide a comprehensive treatise of technologies of importance in breeding of domestic animals but to provide the reader with an appreciation for "what's new" within the framework of the animal breeding industry along with relevant implications. In early chapters, this volume brings together concise summations of the state-of-the art, along with thoughts on future directions, for several animal industries (artificial insemination, embryo transfer, poultry breeding, and aquaculture) of great importance in the production of food. In the middle chapters, developing technologies (sex selection, frozen storage of oocytes and embryos, *in vitro* fertilization and embryo culture, amphibian nuclear transplantation, parthenogenesis, identical twins and cloning in mammals, and gene transfer in mammalian cells) along with potential applications and impacts on animal production are presented. In final chapters, analyses of potential genetic impacts (artificial insemination, sex selection, embryo transfer, cloning, and selfing in dairy cattle) and special economic considerations (benefits of reproductive management, synchronization of estrus, and artificial insemination of beef cattle and sheep) are presented, cases for which such illustrations were deemed most appropriate. The 14 chapters were written by 17 specialists representing a similar number of disciplines, but pulled together by a common interest in the intriguing possibilities for rapid extension of basic knowledge generated through research to very practical applications in animal breeding. Emphasis is on food-producing and large domestic animals, especially the bovine species. Reproductive processes in these animals represent the most appropriate targets for application of the new technologies presented.

As this book goes to press, feelings of sadness are shared with the family and many friends of Dr. J. B. Peters, a fellow contributor and esteemed colleague, recently deceased. Dr. Peters will be remembered for his devotion to the improvement of agriculture and for his good work in animal science at West Virginia University.

The editors gratefully acknowledge the excellence and cooperation of each of the contributors, and the input of every participant in the OTA Conference on "Impacts of Applied Genetics: Animal Breeding," and of scientists far and wide who have generously contributed information to the OTA assessment and to the literature—with apologies in advance for failure, in many instances, to make more proper acknowledgments. Special thanks are due to Miss Pamela J. Salsbury for secretarial assistance and to Mr. Lawrence Burton, Dr. Gretchen S. Kolsrud, and Dr. Zsolt Harsanyi of the Genetics and Population Program, OTA, U.S. Congress, for their effective collaboration in the early stages of this project. The interest, encouragement, and cooperation of the staff of Academic Press in expeditious publication are greatly appreciated.

Benjamin G. Brackett
George E. Seidel, Jr.
Sarah M. Seidel

I

Introduction

1

Perspectives on Animal Breeding

GEORGE E. SEIDEL, JR., AND BENJAMIN G. BRACKETT

I. OVERVIEW OF ANIMALS IN SOCIETY

A. Role of Animals in Human Society

Homo sapiens has evolved in an environment with other animals. He has depended on them for food, hair and hide, power, sport, prestige, and companionship. The degree of dependence varies among cultures and groups within cultures, but the dependency has been absolute in terms of at least one essential

3

NEW TECHNOLOGIES IN ANIMAL BREEDING

nutrient, vitamin B_{12}, which is not found in nature outside of animal tissues. Furthermore, in the United States, animal products provide 69% of the protein in the average human diet, 80% of the calcium, about two-thirds of the phosphorus and vitamin B_2, the majority of vitamin B_6, as well as a substantial minority of most other nutrients (Harper *et al.*, 1980). Clearly, many human societies could survive without animals, but this would be both difficult and artificial, particularly with nonaffluent cultures.

Intangible values associated with man's relationships with animals, especially dogs and cats, often have profound emotional and psychological influences. Such influences have improved the quality of life for many, most strikingly for the very young, for the very old, and for those with certain kinds of mental illness. The following pages deal mainly with large, domestic animals, but food-producing species, for which the application of new technologies to enhance breeding might find encouragement from human society worldwide, will also be discussed. The important role of laboratory animals, e.g., frogs, hamsters, mice, rats, and rabbits, as experimental models to improve efficiency in animal breeding should be appreciated. The use of animals with appropriate genetic characteristics for studying human disease, nutrition, and other health-related problems is also of paramount importance to human society.

B. Competition between People and Animals for Food

It is often less efficient to feed plant products to animals than to consume them directly. On the other hand, the most abundant plant material, cellulose, is of little direct nutritive value to man, whereas cellulose is an excellent feed for ruminants and animals with a functional caecum. Domestic animals frequently eat what is aesthetically objectionable, of marginal or no food value, or even harmful to man, e.g., grass along the road, garbage, acorns, brewers' grains, beet pulp, citrus pulp, straw, cottonseed hulls, and stalk silage. It is true that animals compete with people for grain, but this argument has frequently been presented in a simplistic manner. A small amount of grain (or other relatively high energy substances, including some by-products) can greatly amplify the amount of food produced by animals that eat primarily nongrain feed. Furthermore, when grain becomes scarce and prices increase, much less is fed to animals. This happened with dairy cattle in 1972 (see Fig. 3 in Chapter 2, this volume). Animals, therefore, act as a buffer for grain prices and supplies. When there are surpluses, animals utilize the excess. When there is scarcity, alternate feed supplies are used and/or animal numbers are reduced.

C. Applications of Genetic Improvement

One can logically conclude that domestic animals are likely to continue to be essential in most human societies and that their continued genetic improvement is worthwhile. By and large, this book is limited to considerations of how to apply new technologies to food-producing domestic animals. However, many of the technologies will also be applicable to companion animals, circus and zoo animals, laboratory animals used for agricultural, biomedical, and industrial research, and game and nongame wild animals (including avian and aquatic species). The technologies described may be particularly effective in rescuing some species from extinction or even possibly in recreating extinct species. The technologies also can contribute to greater quantities and improved quality of animal by-products, which frequently are worth more than the food component, e.g., wool, hides, aggressive behavior at rodeos, and velvet from antlers for use as an aphrodisiac in the Orient.

II. PERSPECTIVES ON ANIMAL BREEDING

A. Interactions between Genotype and Environment

Why improve, or at least modify, the genetic makeup of animals? Fortunately, the classical arguments of genotype versus environment are obsolete. It is generally recognized that the environment is extremely important in producing a phenotypic product, such as milk or meat. However, in order to obtain a desired phenotype, both the genotype and the environment must be considered jointly. For example, poultry, regardless of genetic makeup, would freeze to death in winter in much of North America if not protected. Similarly, while dairy cows have calves that can be used for meat, such cows do extremely poorly on the western range in North America because they cannot find enough feed to have normal reproductive cycles and provide milk for the calf too; the genetic propensity to divert nutrients to milk results in a huge nursing calf but no pregnancy for the next year. Conversely, although beef cows produce milk, they would quickly be culled in a dairy herd because of being unprofitable, i.e., they are not genetically geared to turn the large quantities of nutrients into large quantities of milk. It should be noted, however, that, although interactions between genotype and environment are important, they are not so exacting that strains developed in one environment cannot be used to advantage in some others. For example, bulls whose progeny are ranked differently in terms of milk production maintain the same rankings regardless of whether the cows are kept in North America, central Europe, southern Europe, Israel, or subtropical Latin America (see McDowell *et al.,* 1976.)

B. Characteristics of Genetic Improvement

Although it is important to provide optimal environments for animal production, animal breeding is aimed at providing optimal genotypes for the available environments. One extremely important aspect of genetic improvement is its cumulative and permanent nature, at least in environments that do not change drastically over the years. This compound-interest effect is of great economic value. Further, such improvements are the result more of information, thought, and use of computers than additional energy or feed. In some cases, genetic improvement means that more feed is required per animal, but the number of animals required for the same output is usually so reduced that the net feed used per unit of final product is much less. Sometimes there are additional fringe benefits. For example, genetically improved high-producing dairy cows also have fast-growing bull calves, a desirable trait of significant economic importance.

C. Beginnings of Animal Breeding

Before considering new animal breeding technologies, it seems appropriate to briefly review some historical aspects. Animals have been domesticated for millenia, and the process continues today (see Chapter 5, this volume). The choice of species to domesticate is based on many elements, including chance, human needs, biological constraints, and probably an anthropomorphic component too. The successes and failures of the initial phases of domestication should be considered the first work in animal breeding because they set the stage for subsequent improvements. A notable advance was the formation of societies in England that evolved into animal breed societies, primarily in the nineteenth century. These groups kept quantitative records on animal performance and developed the philosophy "breed the best to the best." Breed societies are successful institutions that survive to the present day.

D. Interdisciplinary Aspects of Animal Breeding

Nearly everything that comprises the science of animal breeding springs from the twentieth century, including the theoretical framework and the practical tools. This book concentrates on the physiological tools, such as artifical insemination, estrus synchronization, embryo transfer, and concepts such as cloning, parthenogenesis, and *in vitro* fertilization. Extremely important components of animal breeding that are not dealt with in detail include the more abstract concepts of crossbreeding and hybrid vigor; the concepts of heritability, repeatability, and selection indexes from population genetics; statistical concepts such as linear regression and variance components; mathematical techniques of linear algebra and calculus; and computer technology and language. Few realize that some of the mathematics routinely used in evaluating domestic animals was not thought out until the second half of this century (e.g., Henderson, 1953). The purpose of mentioning these areas is that they are an all-important part of animal breeding, and indeed, they must be used in concert if desired goals are to be realized. Contributions of disciplines concerned with nutrition, animal health, environmental sciences, and marketing are also obviously important.

E. Problems of Animal Breeders

Animal breeding may be defined as the selection of the parents of the next generation. The central principle is that offspring will be more like their parents than unrelated individuals (there are exceptions, e.g., Palomino horses) and that if one selects desirable animals as parents, more desirable offspring will result than with no selection. Two problems arise immediately. First, with conventional sexual reproduction, the offspring receive only a random half of the alleles of each parent, and therefore, the results are not completely predictable, especially with traits of low heritability. Second, there may be little opportunity for selection of females if population size is to be maintained. For example, 70-90% of female dairy calves must be retained in the United States, just to replace dairy cows culled for reasons unrelated to selection, such as old age and disease. Because mammalian females can only carry limited numbers of pregnancies to term, whereas males can produce thousands of progeny through their sperm, there is generally much more opportunity to breed selectively from a few superior males. This creates problems with sex-limited traits, such as milk production, where males must be evaluated indirectly via their female relatives (primarily their daughters).

F. Circumventing the Chance in Sexual Reproduction

Theoretically, there are several ways to overcome the chance inherent in sexual reproduction. The most successful has been inbreeding to various degrees of homozygosity. Inbred animals are generally such poor performers phenotypically that most strains as a rule die out before much homozygosity is attained. If, however, an essentially inbred strain is produced that reproduces fairly successfully, the inbreeding depression in general phenotypic performance can be circumvented by crossing with another inbred strain. Such crosses are frequently superior to the general population, but the real advantage is that the genotype, and to a considerable extent the phenotype, of such crossbred individuals is completely predictable. More than a dozen generations of brother–sister matings are required to produce inbred strains. This is impractical with many farm animal species, especially since most strains die out in the process. Clearly, any procedures that could speed up this process would be invaluable.

A related method of inbreeding is mating an animal to itself. Though impossible in mammals at present, such a result may someday be possible via parthenogenesis. Selfing of males could be accomplished by fertilization of an oocyte with two sperm cells and microsurgical removal of the female pronucleus.

Clearly, the best theoretical method of avoiding the uncertain results of sexual reproduction is through asexual reproduction. Several methods of asexual reproduction have recently been successful with mammals. The simplest of these methods is to divide preimplantation embryos to form identical twins or multiplets. Illmensee and Hoppe (1981) have recently accomplished the same end by transplanting nuclei from embryonic cells. Perhaps it will be possible to make copies of adults some day by similar asexual means. One idea currently under serious consideration is to divide embryos that have the potential to develop into outstanding animals into two halves: one of the halves would be frozen, and the other half would be transferred to a recipient mother. Should the animal resulting from embryo transfer be outstanding, there would be a spare copy in the freezer.

G. Special Opportunities

Animal breeders deal with two classes of resources: (1) genetic material in the form of animals and their gametes and (2) knowledge, skills, and information. These are valuable resources that individuals, agencies, and governments are willing to purchase. Valuable genetic traits include genes for milk production in North American Holsteins, high growth rates in Danish swine, prolificacy in Finnish sheep, agility in Arabian horses, disease resistance in African cattle, and fleece characteristics of Australian sheep. With frozen semen and/or embryos, it is possible to move these genetic resources around the globe easily and inexpensively, with the additional benefit that the possibility of disease transmission is markedly reduced, relative to moving animals. There is also an epigenetic and developmental advantage if an animal is born into the environment in which it is expected to perform and reproduce (Seidel, 1981). Exports of germ plasm from North America alone amount to tens of millions of dollars annually (Rumler, 1981). Similar movement of genetic information within countries is even more important.

The exchange of skills and knowledge is also important. Many scientists have traveled abroad to obtain the latest information. There are thousands of students learning these methods in countries other than their own, and there are dozens of trade agreements that include technical training packages (Nelson, 1981). One extremely important aspect of this activity is that expensive resources, such as energy or certain minerals, are not involved to any great extent, nor is expensive equipment usually crucial to the application of new technologies. The major resource required is people.

A vital aspect of such exchanges of information and genetic material among cultures is increased understanding of one another's problems. Much good will has been exchanged at the same time, and this may be the world's most important commodity.

III. CONCLUSIONS

Animal breeding is excruciatingly slow to show results, especially with species such as cattle, which have long generation intervals. In fact, a cattle breeder may only work with six or seven generations in his lifetime (Seidel, 1981). Nevertheless, there has been phenomenal genetic progress with regard to increased milk production; so much so that the upward genetic trend in the animal population has interfered with the evaluation of the genetic merit of bulls. When steady improvement became a reality, new methods had to be worked out for the genetic evaluation of dairy sires.

Progress is obviously possible, and one can confidently expect the new technologies described in the following pages to contribute to more rapid genetic progress. This is certain to benefit mankind in numerous ways. The intellectual activity involved in developing new technologies and the strategies for applying them appropriately is a very positive human endeavor; this activity, even if not always successful, is worthy of pursuit in its own right. Animal breeders have considerable pride in their accomplishments, and technicians, extension agents, scientists, administrators, politicians, and taxpayers also have a right to acknowledgement of their constructive roles in this important societal endeavor.

A final point concerns the applications of these techniques to man himself. One can certainly concoct fiendish applications, but there are also many opportunities for good. Artificial insemination is being used routinely and *in vitro* fertilization and embryo transfer are beginning to be used. Within a decade, it may be within a man's power to alleviate incalculable human misery by repairing a deleterious gene in his offspring and subsequent generations of descendents. There does not seem to be any more potential for abuse of this technology than for any other. Nevertheless, it is worthwhile to take precautions. This is best accomplished by the acquisition of appropriate knowledge, careful study, engagement in considerable dialogue, and the application of wisdom.

REFERENCES

Harper, A. E., Dwyer, J., Brown, M. L., Allen, C. E., Bergen, W. G., Dierks, R., Guthrie, H. A., Hegsted, M., Henderson, L. M., Jacobson, N. L., Kellough, M., Leveille, G. A., Nesheim, M. C., Simopoulos, A., and Speckman, E. (1980). *In* "Animal Agriculture" (W. G. Pond, R. A. Merkel, L. D. McGilliard, and V. J. Rhodes, eds.), p. 36. Westview Press, Boulder, Colorado.

Henderson, C. R. (1953). *Biometrics* **9**, 226–252.

Illmensee, K., and Hoppe, P. C. (1981). *Cell* **23**, 9–18.

McDowell, R. E., Wiggans, G. R., Camoens, J. K., Van Vleck, L. D., and St. Louis, D. G. (1976). *J. Dairy Sci.* **59**, 298–304.

Nelson, R. E. (1981). *Proc. 7th IETS Owners-Managers Workshop, Denver 1981,* pp. 31–42.

Rumler, R. H. (1981). *Holstein World* **78**, 1016–1026.

Seidel, G. E., Jr. (1981). *Science (Washington, D.C.)* **211**, 351–358.

II

Animal Industries Heavily Dependent on Reproductive Technology

2

The Artificial Insemination Industry

ROBERT H. FOOTE

13

NEW TECHNOLOGIES IN ANIMAL BREEDING

I. HISTORICAL PERSPECTIVE

When the technique of artificial insemination (AI) was first introduced into the United States from Denmark in 1938, who would have predicted the tremendous impact that AI would have on the improvement of dairy cattle? Never, in recorded history, has the genetic improvement of animals been so great as has occurred in dairy cattle in the past 25 years.

Initially in Europe the technique had been used more as a means of attempting to control venereal disease. In the United States, the goal primarily was one of genetic improvement, but the initial struggle was simply to get cows pregnant in order to have the dairymen accept the AI approach. The technique of depositing the semen in the reproductive organs of the cow had been developed in Denmark. But there was the need to train technicians to become skillful in this art and to educate dairymen to detect estrus and to report cows for insemination at the right time for conception. Equally important was the Herculean task of learning how to harvest fertile spermatozoa in large numbers and to maintain this fertility on the long journey from the bull stud to the farm for use over a period of a few days. This was quite a contrast to natural service, where the transfer of spermatozoa is a matter of seconds or less, and sperm cells are not exposed to the "light of day." Some of these early events have been described by Perry (1968).

AI was promoted primarily to take advantage of the fact that each bull produced enough sperm cells to be used to inseminate many cows. It was thought that bulls with superior genetic merit could be accurately identified for AI service. Later it was found that, with the information available in the beginning, bull selection was ineffective. Fortunately, when this ineffectiveness was realized, AI was well established, and steps were undertaken to develop a highly successful program of young sire selection, testing, and culling and to obtain wide-spread use of sires that survived this rigorous program (Henderson, 1973).

Success resulted from a combination of the vision of a few farsighted extension personnel who saw the need, of researchers in physiology and genetics who had basic tools to research the problems, of the development of record systems to monitor results in AI and of the use of production records through the university-organized farmer-run cooperative cow-testing associations (now, dairy herd improvement associations). It is a lasting tribute to the land grant system that its member colleges continuously strive to find better ways of effectively feeding a hungry world.

Dairy cattle will be used as a prime example throughout because the state of AI is more highly developed in cattle, and the application has been much greater in dairy cattle than in other livestock. The basic principles usually apply to other species. Some details for beef cattle are given in Chapter 13 of this volume.

II. ADVANTAGES OF ARTIFICIAL INSEMINATION

There are many advantages of AI, and few disadvantages. The latter are associated with such conditions as great distances and poor communications where AI service often is not available or is uneconomic.

The greatest advantage is the genetic improvement possible for quantitative traits through intensive sire selection. Also, there has been a reduction in the frequency of recessive lethal genes (Foote *et al.*, 1956). Second, the control of certain diseases, especially venereal diseases, has been a major accomplishment (Bartlett, 1980). The goal of the bull stud members of the National Association of Animal Breeders is to breed animals that are specific pathogen-free. Third, is the economical service. With reasonably priced semen, the value of progeny produced exceeds the costs of semen and service (Everett, 1975). A fourth advantage is safety through elimination of dangerous males on the farm. Considering this in a lighter vein, no inseminator has ever been reported to butt a dairyman. Another advantage is the more complete and accurate records generally kept when AI is used. These are necessary for good reproductive management of the herd and, when reported to central stations for computer processing, they provide unique information of extraordinary value for fertility and genetic research. Without these records made available to researchers, progress would have been much slower. Finally, the successful AI technique and program has provided the essential prerequisites for the successful use of the advanced breeding techniques of (1) estrous cycle regulation and (2) embryo transfer. (For a detailed account of AI in dairy cattle, see Salisbury *et al.*, 1978; for a recent overview of several species, see Foote, 1980.)

III. GROWTH OF ARTIFICIAL INSEMINATION

A. Cattle

The growth of AI in dairy and beef cattle is summarized in Table I. Comparable figures since 1972 are more difficult to obtain because of the widespread sale of frozen semen and direct owner use. The total number of dairy heifers and cows inseminated in 1979 was about 7,000,000 in a national dairy herd of approximately 12,000,000 dairy cows and heifers of breeding age. The trend has been to consolidate a large number of bull studs formed in the early years into a few large AI organizations. Most of the dairy cows in the United States are inseminated with semen produced by about 10 organizations. The number of breeding units of semen harvested per sire per year has increased. Herd size has increased but, the number of herds has declined rapidly with a declining dairy cow population.

TABLE I

Growth of Dairy and Beef Artificial Breeding Programs in the United States[a]

Year	Studs	Sires in service Dairy	Beef	Total	Average per stud	Herds	Dairy cows bred to Dairy bulls	Beef bulls	Beef cows bred to beef bulls	Total cattle bred	Cows bred per sire
						Number					
1939	7	—	—	33	4.7	646	—	—	—	7,359	228
1940	25	—	—	138	5.5	2,971	—	—	—	33,977	246
1941	35	—	—	237	6.8	5,997	—	—	—	70,751	299
1942	46	—	—	412	9.0	12,118	—	—	—	112,788	274
1943	59	—	—	574	9.7	23,448	—	—	—	182,524	318
1944	56	—	—	657	11.7	28,627	—	—	—	218,070	332
1945	67	—	—	729	10.9	43,998	—	—	—	360,732	495
1946	78	—	—	900	11.5	73,293	—	—	—	537,376	597
1947	84	—	—	1,453	17.3	140,571	—	—	—	1,184,168	815
1948	91	—	—	1,745	19.2	224,493	—	—	—	1,713,581	982
1949	90	—	—	1,940	21.6	316,177	—	—	—	2,091,175	1,078
1950	97	—	—	2,104	21.7	409,300	—	—	—	2,619,555	1,245
1951	94	—	—	2,187	23.3	548,300	—	—	—	3,509,573	1,605

16

Year											
1952	94	—	—	2,324	24.7	671,100	—	—	—	4,295,243	1,848
1953	96	—	—	2,598	27.1	755,000	—	—	—	4,845,222	1,865
1954	93	—	—	2,661	28.6	805,000	—	—	—	5,155,240	1,937
1955	79	—	—	2,450	31.0	845,900	—	—	—	5,413,874	2,210
1956	79	—	—	2,553	32.3	900,400	—	—	—	5,762,656	2,257
1957	75	—	—	2,651	35.3	946,000	—	—	—	6,055,982	2,284
1958	71	—	—	2,676	37.7	975,372	—	—	—	6,645,568	2,483
1959	64	—	—	2,460	38.4	930,059	—	—	—	6,932,294	2,816
1960	62	—	—	2,544	41.0	910,000	—	—	—	7,144,679	2,808
1961	56	—	—	2,486	44.4	863,781	7,047,148	435,592[b]	—	7,482,740	3,010
1962	56	2,036	420	2,456	43.9	862,150	6,837,681	911,006[b]	—	7,748,687	3,155
1963	51	2,158	401	2,559	50.2	621,141	6,468,545	969,748	235,289	7,673,582	2,999
1964	50	2,140	398	2,538	50.8	654,311	6,165,599	1,117,395	464,959	7,747,953	3,053
1965	46	1,867	449	2,316	50.3	591,859	6,301,178	963,657	615,147	7,879,982	3,402
1966	44	1,949	439	2,388	54.3	540,265	6,413,453	873,127	695,181	7,981,761	3,342
1967	35	2,012	364	2,376	67.9	458,782	6,259,425	788,933	672,819	7,871,265	3,313
1968	33	2,028	352	2,380	72.1	407,375	6,423,786	714,850	795,242	7,933,878	3,334
1969	30	1,955	390	2,345	78.2	387,979	6,590,147	694,916	924,381	8,209,444	3,501
1970	31	1,911	364	2,275	73.4	369,197	6,693,216	615,322	1,258,446	8,566,984	3,641
1971	24	1,958	349	2,307	96.1	350,611	6,759,215	525,956	1,357,918	8,643,089	3,620
1972	26	2,167	347	2,514	96.7	—	—	—	—	—	—

[a] From Dairy Herd Improvement Letter, ARS-NE-1, 48(4):10. June/July 1972. Published by ARS, USDA.
[b] Probably includes some beef-to-beef inseminations.

Beef cattle AI has never been a major factor in the production of beef. The proportion of beef animals inseminated has never exceeded 5%, and it currently is only 3%. While the technology of semen handling and insemination is similar in beef and dairy cattle, range management of much of the national beef herd makes AI impractical for these animals (see Chapter 13, this volume).

B. Other Livestock

AI has been used on a limited basis for many years in horses, sheep, goats, and swine Segments of the horse industry are opposed to AI and do not allow it in some breeds, if animals are to be registered. Detection of estrus is a problem in mares as is survival of equine spermatozoa. However, AI is used as an adjunct to natural service on many large horse breeding farms.

Sheep breeders find the price of good rams low enough, and the convenience of rams managing reproduction in their flocks important enough so that AI is rarely used in North America. The goat industry is small, but oriented toward artificial insemination. This offers the possibility of both improving the milk production and eliminating the strong buck odors from the premises of many small goat herds.

Swine AI has not developed in the United States to the extent that it has in many other countries. Most United States swine breeders find the boar more convenient to use than AI. In concentrated swine producing areas of the Midwest, a small proportion of the sows are inseminated artificially. Professional AI service is not available, so each breeder must learn the technique and reserve time for AI. Because the value of each sow is low (relative to a cow or mare), the cost per service must be low to be economic.

C. Poultry

Poultry breeding is in the hands of a few large companies. Considerable use of AI is made in controlled breeding (Lake, 1980). Fresh semen can be collected rapidly, and a large number of females can be inseminated per hour. Hens need to be inseminated only about every 2 weeks. The entire turkey industry depends upon artificial insemination. The selection for extensive development of the breast (white meat) has produced birds that cannot copulate readily. AI with fresh semen is used almost exclusively.

IV. MAJOR TECHNOLOGICAL MILESTONES IN ARTIFICIAL INSEMINATION DEVELOPMENT

A. Harvesting Fertile Spermatozoa

Initially, little was known about the sexual behavior of bulls, how frequently semen should be collected, and other factors of management that could affect the yield and harvest of spermatozoa (Salisbury *et al.,* 1978). The importance of this harvest is that the number of fertile spermatozoa that can be harvested from a bull is directly proportional to the potential number of progeny such a male can sire. In the early days of artificial insemination, the number of sperm cells harvested per sire per week was about 8-10 billion. Through the study of sexual behavior, methods of sexual preparation have been developed, along with an optimal schedule of semen collection, which have resulted in harvesting 30-40 billion spermatozoa per sire per week.

B. Keeping Stored Spermatozoa Fertile

To sustain the life and fertility of spermatozoa, it was necessary to find media that could provide nutrients and protection during refrigeration. The latter was necessary to slow down the very active metabolism of spermatozoa and to prolong their life span. A successful egg yolk–citrate medium was developed, which permitted semen to be carried in ice chests in the field (~5°C) for several days before fertility decreased considerably. With this medium, about 40-50% of the cows conceived with one insemination. Soon it was discovered that there were infectious organisms in semen from clinically healthy bulls that could be controlled by antibiotic treatment of the semen. This treatment improved fertility by 5-10%. Other improvements in extenders were made, and by the early 1960s the reported fertility in AI had reached unprecedented levels of 70% or higher (Foote, 1975, 1980; Graham, 1978; Salisbury *et al.,* 1978). Fertility of spermatozoa can be preserved for years by proper freezing (Foote, 1978), although damage may occur under field handling conditions (Salisbury and Lodge, 1978).

C. Optimal Number of Spermatozoa per Insemination

From initial research it was found that diluting semen with various physiological buffers was harmful. There was a "dilution effect." However, with egg

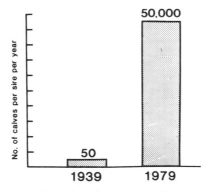

Fig. 1.   Change in the potential number of progeny per sire per year from 1939 to 1979 as a result of artificial insemination.

yolk–citrate and more complex extenders, the life span of the sperm cell could be prolonged by the protective media. Then it was possible to increase the extension rate from 1 part of semen to 2 parts of extender to 1:10, 1:25, 1:50, 1:100, and to even higher rates (Salisbury *et al.,* 1978), depending on the concentration of spermatozoa in the original ejaculate. Several hundred cows could be inseminated with the extended semen from each ejaculate. An example of the dramatic effect of AI on the number of progeny that could be produced per sire per year is illustrated in Fig. 1. The possibility of replacing each 1000 bulls with one superior bull has enormous implications, which will be alluded to at various times as the AI record unfolds. The genetic implication of extensive testing and intensive culling is the most obvious one.

D. Frozen Semen

The discovery of how to preserve mammalian cells by freezing with a cryoprotective agent, such as glycerol, resulted from studies on freezing bull semen. This has had widespread application in preservation of blood and other lifesaving materials. Actually, freezing spermatozoa successfully has not been a key to the development of AI. However, because of its convenience, freezing is an essential ingredient, and it permits semen collected at any time and place to be used at almost any other subsequent time and place.

Freezing of semen kills up to one-half of the sperm cells. Consequently, more sperm cells must be provided initially per breeding unit to have enough cells survive to give good fertility. But because all semen collected and frozen can be stored until used, the loss during freezing is offset by the fact that little need be discarded. This is in contrast to the short storage life of liquid semen used previously. The net result is that Fig. 1 is appropriate for frozen semen also.

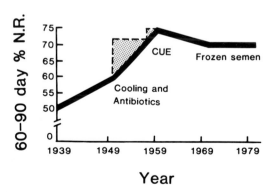

Fig. 2. A summary of fertility (based on nonreturn rates) during the past 40 years of breeding dairy cattle in New York State. Improvements in liquid semen processing and extension (CUE extender) improved fertility during the first 25 years.

Freezing semen tends to lower the fertility of surviving cells slightly. This is shown in Fig. 2, along with fertility changes prior to the widespread adoption of frozen semen in the 1960s. The decrease in fertility is considered to be offset by the greater convenience of frozen semen. This point is stressed because convenience is of great importance to acceptance. Even the best technology will not find widespread application if it is not convenient.

E. Trained Personnel

Perhaps the most important component that has allowed semen to be extended further, frozen, and deposited successfully in larger herds is the development of aids (Foote *et al.*, 1979) and of trained personnel (Pickett *et al.*, 1976; Berndtson and Pickett, 1978; Saacke, 1978). This is an enormous investment. There is ample evidence in the AI industry that the application of science is really an art that succeeds only in the minds and hands of the skillful. Innate ability, rigorous training, continuous dedication, and practice all are important. These are also the ingredients that lie behind the success story of American agriculture.

V. IMPACT OF ARTIFICIAL INSEMINATION ON MILK PRODUCTION

A. Physiology and Genetics Related

The impact that a dairy sire can have on the dairy population is a product of physiology and genetics. This is illustrated in the following two equations:

$$\begin{pmatrix} \text{Genetic} \\ \text{impact} \\ \text{per sire} \end{pmatrix} = \begin{pmatrix} \text{Genetic} \\ \text{superiority} \\ \text{of the sire} \end{pmatrix} \times \begin{pmatrix} \text{Number of} \\ \text{progeny} \\ \text{per sire} \end{pmatrix} \quad (1)$$

$$\begin{pmatrix} \text{Number of} \\ \text{progeny} \\ \text{per sire} \end{pmatrix} = \begin{pmatrix} \dfrac{\text{Number of sperm}}{\text{produced/sire}} \\ \dfrac{}{\text{Number of sperm}} \\ \text{required/cow} \end{pmatrix} \times \begin{pmatrix} \text{Fertility} \\ \text{of the} \\ \text{semen} \end{pmatrix} \times \begin{pmatrix} \text{Fraction of} \\ \text{the semen} \\ \text{used for} \\ \text{insemination} \end{pmatrix} \quad (2)$$

All of the improvements made in sire testing will affect the sire's genetic superiority. All the improvements in harvesting sperm from the bull, preserving them with minimal loss, and skillfully placing the right number in the cow at the proper time will affect the number of progeny. For Eq. (2), assume the following:

1. Sperm Cells harvested per sire per year: $1{,}500{,}000 \times 10^6$
2. Number of sperm cells required per cow: 15×10^6

3. Fertility: 50% (0.50)
4. Semen used: 100%

Solving:

$$1{,}500{,}000 \times 10^6/15 \times 10^6 \times 0.50 \times 1 = 50{,}000 \text{ progeny}$$

Many AI organizations use more than 15×10^6 sperm per breeding unit. Then the number of progeny would average less than 50,000 per sire.

B. Genetic Improvement

Now consider Eq. (1). Little progress was made during the first decade of using AI. The number of progeny possible per sire in the beginning was small. Also, since milk production is a sex-limited trait, records on female relatives were needed for evaluation of sires. Unfortunately, the records of relatives usually were within one herd and were completely confounded with management and other environmental factors. Also, sample size was small.

Artificial breeding provided a breakthrough. Cows were milked in many herds. Fortunately, some of these herds were on a milk testing program (Dairy Herd Improvement). At this time, simple computer technology was developing and scientists with excellent training in quantitative genetics were available to make use of these records. Techniques to estimate the breeding value of sires with accuracy were developed (Henderson, 1973). Henderson (1973) has reviewed optimal programs developed to take into account:

1. Pedigree selection of young bulls.
2. Number of bulls to sample relative to (a) the number of bulls to save after sampling, (b) the accuracy of evaluation with varying numbers of tested daughters, and (c) the proportion of the cow population to breed to tested versus untested sires.

Adoption of these recommendations in the early 1950s began to show positive results in the late 1950s (Fig. 3). *For each $1.00 invested in research, it is estimated that this program has returned $50.00 to the economy.*

The genetic improvement is permanent and makes up about 30% of the total increase. The balance is due to feeding, housing, and management, which has to be supplied each year. In Fig. 3, the fluctuating environmental influence is apparent. In terms of the annual effect of the economy, consider the following facts.

1. About 7 million dairy cows and heifers are inseminated annually.
2. Genetic superiority is about 1000 lbs (454 kg) of milk.
3. Milk price is $\geq$ $.10/pound ($\geq$ $.22 cents/kg).

Gross annual value of the genetic component $= 7.0 \times 10^6 \times 1000 \times 0.10$
$$= \$70{,}000{,}000$$

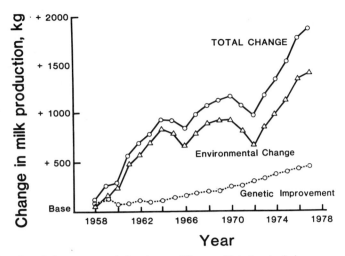

Fig. 3. Genetic improvement during the past 20 years. Note the steady improvement, which is permanent, whereas the environment has fluctuated.

The shift in the gene frequencies through selection, which is responsible for this increase, should be permanent in a randomly mating population.

Genetic gain should increase more rapidly in the near future when the impact of present testing programs is felt. Pedigree selection of young bulls is improving. With more progeny per sire, fewer need to be saved to supply semen. The intensity of selection is also increasing as a result of increasing the number of bulls being progeny-tested per bull saved for extensive use following the progeny test. There is no evidence yet that genetic variability is decreasing or that a plateau has been reached. The value of AI and other technologies is discussed in Chapter 12, this volume).

VI. PRESENT SITUATION

A. State of the Science and the Art of Artificial Insemination

The following is a brief overview of the state of the science and the art of AI in several species. The points discussed provide not only a synopsis of the present, but are germane to prognoses concerning the likelihood that changes should and will be forthcoming in the application of AI to these species in the future.

1. Dairy Cattle

It is clear that AI has been enormously successful in improving the genetic merit of dairy cattle and in controlling certain diseases. Dairy cattle lend them-

Robert H. Foote

TABLE II

Technical Feasibility of Using AI

Species	Fertility of Semen		Any limitations[a]
	Liquid	Frozen	
Dairy cattle	Good	Good	Requires good detection of estrus, but this is possible with well-confined milking cows.
Beef cattle	Good	Good	Animals must be restricted for detection of estrus and insemination. Range is unsatisfactory.
Sheep	Good	Fair	Problem on range. Low value/ewe, so AI costs must be very low.
Goats	Good	Fair	Detection of estrus in small herds. Insemination is more difficult. Availability of semen.
Swine	Good	Fair	Detection of estrus. Slightly smaller litters with frozen semen. Cost/sow must be kept low.
Horses	Good	Fair	Long estrus. Multiple inseminations needed. Frozen semen conception rates are low.
Turkeys, chickens	Good	Poor	None in breeding flocks. Use of fresh semen is method of choice.

[a] Note that cattle have an advantage over other species in number of progeny possible per tested sire per year. Cattle, 50,000; sheep and goats, 5,000; swine, 2000; horses, 750. The successful development of frozen semen to bank would be especially advantageous in seasonal breeders as sheep, goats, and horses.

selves to AI because of the confined housing, regular contact with the dairyman, high value, and generally good results from the application of all phases of semen technology, including freezing of semen (Foote, 1978; Pickett and Berndtson, 1974). In Table II the feasibility of widespread use of AI in several species is compared.

2. Beef Cattle

This is the largest animal industry in terms of numbers and direct dollar value of the animals. The technology of semen collection, processing, and insemination is the same as for dairy cattle. Beef performance testing and progeny testing are extensive, although not as well developed as for the dairy industry. Crossbreeding is common in beef cattle. The practice of using AI eliminates the need for keeping more than one line of animals on a ranch, because the females can be crossed with another breed of sire through the use of frozen semen.

With beef cattle, the large number of cows that are scattered widely on the range presents a problem. Detection of estrus is poor unless cows for breeding are brought into more confined areas. When this is done, an effective method of estrous cycle regulation can be applied. With proper management, these two techniques can be combined to breed beef cattle effectively. These techniques could also facilitate breeding more dairy heifers artificially. The pros and cons and economics are discussed in detail in Chapter 13 of this volume.

3. Sheep

AI of sheep is possible from the standpoint of technology, but practically it has not been a success (Gustafsson, 1978; Chapter 13, this volume). The labor costs per value of animal inseminated, problems of handling animals under range conditions, detection of estrus, and the relatively low cost of performance-tested rams available for natural service have outbid the AI program in the market place. However, under intensive management, with improved semen freezing, and with estrous cycle regulation, AI could have limited application.

4. Goats

Technically AI is feasible in goats also. Goats are housed in confined areas, but the industry is small. The value per animal, labor costs, and the problem of detection of estrus limit the potential application of AI. However, there is great interest by goat breeders in selective breeding by AI. Some prefer the smell of clover to bucks, and AI would be of special advantage to the hobbyist or 4-H youth who may live on the edge of populated areas.

5. Swine

Until recently, the fertility of frozen swine semen has been low, and fertility still is substantially lower than with liquid semen (Larsson, 1978; Pursel, 1979). Litter size has tended to be somewhat smaller when frozen sperm was used. Detection of estrus and the small number of sows that can be inseminated (footnote to Table II) have limited the application of AI. Also, tested boars are available at relatively low cost. However, with intensive housing increasing in the industry, AI becomes more attractive. Sows can be synchronized through weaning litters simultaneously. A method needs to be developed for gilts. More intensive selection of boars can be practiced, and the hazards of handling boars can be minimized by AI. A good method of detection of estrus in sows confined in stalls without exposure to boars is needed for successful AI.

6. Horses

Frozen horse semen has not produced high fertility under practical conditions (Sullivan, 1978; Rowlands and Allen, 1979). Most artificial insemination is limited to the use of fresh semen in certain breeds on large breeding farms. Some breed organizations restrict the use of AI. Commercial attempts to offer AI service have not been successful. Detection of the time of ovulation in the variable estrus period of the mare is a problem. Multiple inseminations during estrus are recommended.

7. Poultry

Use of fresh poultry semen is feasible in the large companies controlling most of the chicken and turkey breeding. Semen is easily collected in proximity to the females, and many females can be rapidly inseminated every 1–2 weeks. Frozen semen has not given equivalent fertility (Lake, 1980).

B. Worldwide Overview

It is difficult to obtain accurate statistics from some countries. Table III was based on Iritani (1980) and other literature. Data are not included from China, where millions of animals are inseminated on collective farms. Data are sparse from several eastern European countries, where the use of AI is facilitated by the collective farm arrangements. Thus, values in Table III are underestimated, but they do give an indication of the extent of AI and the application of frozen semen.

The cattle industry has adopted frozen semen almost exclusively. This has resulted in substantial international sales of semen (Bartlett, 1980). Sheep are inseminated extensively in eastern Europe and in some South American countries. Frozen semen can be used, but most of the inseminations are with fresh semen. Frozen goat semen is largely experimental, although in France 15,000 goats are inseminated annually in a commercial enterprise. A few sows (9000) are inseminated with frozen semen in the United States. The proportion of sows inseminated is high in Denmark and the Netherlands, where liquid (unfrozen)

TABLE III

Estimated Number of Females Inseminated in 1977 and the Use of Frozen Semen Worldwide[a]

Species	No. of females inseminated	Frozen semen use (%)	Comments
Cattle	>90,000,000	>95	Excludes buffaloes[b]
Sheep	>50,000,000	Experimental	Extensive in eastern Europe
Goats	?	Experimental	Commercial in France
Swine	6,000,000	2	
Horses	Low	Experimental	

[a] Summarized from the literature, especially Iritani (1980).
[b] AI is used extensively in India on 40,000,000 buffalo cows.

semen primarily is used. Use of AI in horses also is limited, except possibly in countries such as China.

The poultry industry is not included in Table III. Nearly all of the turkeys produced in the United States result from hens inseminated with freshly collected semen from turkey toms. AI is widely used in breeding flocks of chickens.

C. Artificial Insemination of Animals Producing Food in a Hungry World

1. General

It is clear that in the future, mankind's need for food will not diminish. There will be competition between humans and livestock for food and space. Regarding space, we see more intensive operations in dairy cattle, beef cattle, and swine. In some parts of the world, swine are primarily scavengers and convert wastes into edible food.

Ruminants are particularly effective in harvesting roughages. Roughages require less fossil fuel to grow than do cereal grains, making the ruminant an efficient utilizer of fossil fuel per unit of foodstuff produced (Reid and White, 1978; Pimentel et al., 1980). Furthermore, when roughages are supplemented with concentrates (grains and various by-products) ruminants often produce more than a pound of edible food per pound of extra feed consumed. In addition, ruminants often serve as the highest bidder for surplus grains. If the United States limited grain production so there was not any surplus, now partially taken care of by heavier concentrate feeding in years with bumper crops, there would be years of potential starvation when plant yields were subnormal.

Some of these relationships are summarized in Tables IV and V. In Table IV, it can be seen that dairy cattle, by virtue of their milk production, are efficient producers of protein per megacalorie (Mcal) of digestible energy consumed. Beef

TABLE IV

Efficiency of Protein Production by Animals under Average Management Conditions.[a]

Food product	Protein: gm/Mcal of digestible energy
Broilers	13.7
Milk (dairy cattle)	12.8
Eggs	10.1
Pork	6.1
Beef	3.2

[a] Adapted from Reid and White (1978).

TABLE V

Energy Output When Feeds Are Utilized; Expressed in Terms of Fossil Energy Input for Production[a]

Kind of feed	Energy output/fossil energy input	
	Gross (Mcal)	Digestible (Mcal)
Soybeans	2.3	2.0
Corn silage	5.9	4.1
Hay	12.5	7.5
Pasture	60–150	40–100

[a] Adapted from Reid and White (1978).

cattle are not as efficient. However, beef cattle can make much of their growth (meat) by harvesting range grasses. The data in Table V indicate that energy output from pasture per unit of fossil energy input is extremely high. To put it another way, ruminants have the digestive ability to convert energy of the sun, which is harvested through photosynthesis by crops requiring little planting, cultivation, and harvesting. Therefore, little fossil fuel is required.

2. Dairy Cattle

It is instructive to examine the dairy industry in the United States in more detail to appreciate the impact that the changes associated with AI have brought about relative to availability of plant and animal products. Total milk production has been fairly stable for many years (Fig. 4). This has resulted from a major increase in milk production per cow and a reduction of cow numbers to about

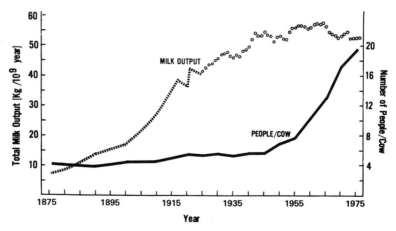

Fig. 4. The increase in total milk production and the increased number of people per cow in the United States. (From Reid, 1978.)

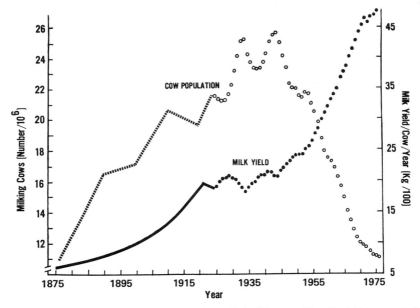

Fig. 5. Trends in the dairy cow population in the United States and in milk yield per cow during the past century. (From Reid, 1978.)

one-half of what they were when AI began (Fig. 5). As a result, one-half as many cows maintain the total national milk supply. That is, more consumers' needs are taken care of by the milk from each cow (Fig. 4). This change also is reflected in fewer dairymen. Three major points emerge:

1. This increase in milk production per cow decreases labor costs and maintenance requirements by the national dairy herd.

2. The lower maintenance requirements, because of fewer cows today, is a saving equivalent to more than 1,000,000,000 bushels of corn annually.

3. Fewer bulls are required for service so more bulls or steers can be raised for meat production. Bulls are raised extensively in Europe, but not in the United States because of unfounded discrimination against meat from young bulls.

VII. FUTURE POTENTIAL: IMPLICATIONS AND APPLICATIONS

A. Dairy Cattle

1. General

As noted, total milk production has been relatively stable for several years. Milk production per cow has gone steadily upward, but cow numbers have gone down proportionately during the past 35 years (Figs. 4, 5).

Production information is given in Table VI on a per cow basis. The number of dairy cows that calved as of January 1, 1980 was 10,810,000. The number has stayed rather stable for the past year, but this number may decrease to less than 10 million in the next decade, if milk production per cow increases further.

An alternative will be that dairy cows will be fed and pastured on lower energy rations consisting mostly of forage. The milk production will be limited by this lower level of energy intake. Then about 5400 kg of milk per year production would be the upper limit. This would require an increase in the number of cows to produce the same total supply of milk. However, in view of the importance of milk in the economy and in the diet, substantial amounts of concentrates probably will be fed as a desireable part of this enterprise, which is labor intensive and requires a large investment in equipment. If this assumption is correct, total dairy cow numbers will follow the current trend of a slight decrease with increasing national product.

Current statistics are summarized in Table VII. Cow numbers, temporarily, are up about 1%, reflecting favorable milk prices. The inventory of frozen semen in storage probably is up slightly. This inventory and resultant sales likely will remain nearly constant, using current technology. The number of breeding units sold is considerably in excess of the number of lactating dairy cows inseminated because (1) many cows require more than one service to become pregnant, (2) some semen is discarded eventually without being used, and (3) some pregnant cows are slaughtered or die. The total number of dairy cows inseminated is about 7,000,000.

Herd size is increasing, and the number of herds continues to decrease. With greater distances between herds, there has been a lower proportion of cows inseminated by professional inseminators. More cows are inseminated by the herd owner or herdsman trained to do his own inseminating. This trend will

TABLE VI

Milk Production per Cow during the Past 25 Years

Year	National average		DHIA (av.)		Non-DHIA (av.)	
	lb.	kg.	lb.	kg.	lb.	kg.
1950	5,300	2,415	9,170	4,168	5,110	2,322
1955	5,840	2,655	9,500	4,318	5,600	2,545
1960	7,030	3,195	10,560	4,800	6,640	3,018
1965	8,300	3,772	11,980	5,445	7,710	3,504
1970	9,750	4,431	12,750	5,795	9,100	4,136
1975	10,354	4,706	13,163	5,983	9,477	4,307
1979[a]	11,500	5,227	15,000	6,818	10,800	4,909

[a] 1979 estimated.

TABLE VII

Current Statistics on Dairy Cows and Breeding Units of Frozen Dairy Bull Semen Produced by United States

Item	Number	Increase from previous year (%)
Dairy cows, Jan. 1, 1980	10,810,000	1.0
Domestic sales of dairy bull semen, 1979	12,467,351	5.1
Export sales of dairy bull semen, 1979	1,836,000	12.6
Custom freezing of dairy bull semen, 1979	681,563	16.0

continue, but should have little effect on statistics of cattle inseminated. Fertility may be affected, but there are no good estimates. The highly experienced professional inseminator potentially can impregnate a higher proportion of cows than the "do-it-yourselfer." However, this aspect tends to be offset by the convenience of the inseminator being a farm employee and potentially able to better coordinate timing of the insemination with ovulation.

2. Economics

The cost of providing professional insemination service is about $5.00 per cow. This covers the labor, car, and depreciation on equipment. The purchase price of semen must be added to that. This varies from a minimum of $3.00 per breeding unit of semen to a few breeding units being merchandized for a few hundred dollars per breeding unit. Overall, a reasonable estimate of the cost of semen is $8.00, as charges by major organizations average from $6.00 to $10.00. Rates have been increasing, as costs cannot be absorbed by increases in business.

The cost of insemination, when done by personnel from the dairy farm unit, is similar, excepting for a few extraordinarily large farms. While there is no travel cost to the farm, the cost of liquid nitrogen, storing fewer breeding units, buying inseminating equipment in smaller quantities, and time required when other operations need attention offset the potential savings. So insemination by personnel on the farm provides a means of using semen from selected sires through AI, where convenient, professional service is not available.

What is the economic value of purchased semen? Everett (1975) and others have estimated the economic returns from semen, taking into account the following:

1. Different levels of genetic merit
2. Costs of service
3. Conception rate
4. Milk price
5. Calf management (calf mortality)

6. Discount rate
7. Years of discounting

The value of semen within a herd is about $21.00 per milking heifer for each 100 kg of a sire's superior genetic merit for milk production. After subtracting the various costs, the net value was calculated to be $3.30 per first service. This was the same for professional and direct on-the-farm service.

3. Changes in Semen Processing

No major changes are forseen. Freeze-dried semen is not likely to be successful enough to use. AI organizations will be making every effort to harvest more sperm cells from bulls and preserve the maximum number through the freezing of semen. Conception rate is so important to milk production that more attention will be given to selecting sires of high fertility (Oltenacu *et al.*, 1980).

4. Banking Semen

Progress should be made in banking semen from young bulls, storing a bank of semen, and discarding many bulls. This will reduce costs and be a hedge against inflation. The following facts are clearly established:

1. Bulls at 2–6 years of age produce large amounts of high quality semen, and they are more fertile as a group than they will be as old bulls.

2. Bulls after 2 years of age are producing large numbers of spermatozoa. With bulls scheduled for semen collection under optimal conditions, 25,000 to 50,000 breeding units per bull can be processed per year.

3. Bull semen stored properly under liquid nitrogen retains its initial fertility for many years.

4. Under almost any set of prices imaginable, it is cheaper to bank and store large quantities of semen from young bulls and slaughter the bulls than it is to keep them for 6 years to obtain a progeny test and then intensively start collecting the selected few bulls to meet breeding requirements.

5. Banking semen also provides some insurance against loss of bulls through natural causes, fire, disease, or other catastrophic accidents.

Clearly, bulls of the less populous breeds, where the number of inseminations is severely limited by the size of the cow population, should have semen banked at an early age. Once the desired number of breeding units is banked the bulls could be slaughtered. Normally, banking can be completed in about 1 year (Coulter and Foote, 1974). This procedure is followed in several European countries. In the United States, advances in the technology of preserving spermatozoa through freezing have been disregarded relative to the possibility of banking semen extensively.

So, it is likely that in the future some of the increased costs of housing and feeding bulls will be offset by this procedure of storing semen rather than housing bulls. At present costs this would be a saving of about $5000 per bull.

5. Sexing Semen

If it were possible to separate X and Y types of spermatozoa without losing many sperm cells in the process and without sacrificing fertility, sexing could have some effect on the AI industry. Van Vleck has discussed the genetic aspects (see Chapter 12, this volume). Also, Foote and Miller (1971) have considered genetic and economic aspects and their advantages and limitations.

If sexed semen does become available, it is likely that more dairy cows would be inseminated artificially. The extra cost of sexed semen should not be more than $10.00 over the price of regular semen. This is based upon a study by Van Vleck and Everett (1976). Sexed semen would be used for insemination of the best cows to produce female replacements primarily. This could result in breeding the remaining part of the herd naturally (Foote and Miller, 1971). However, introduction of bulls presents some management problems. It is estimated that semen sexed by sperm treatment might increase the net number of dairy cows inseminated artificially by 500,000.

6. Embryo Transfer

This will have little effect on semen required. The number of embryos transferred will be small in comparison to the number of cows inseminated in routine AI programs. Also, donors of embryos usually are inseminated with several breeding units (straws or ampules), so the number of breeding units required to produce each offspring is not changed greatly.

7. Estrous Cycle Regulation

The technique of synchronizing estrus, or controlling the time of ovulation, so that cows may be inseminated at a preset time is feasible. With the approval by FDA of prostaglandin $F_2\alpha$ for use in dairy heifers, the number of virgin dairy heifers inseminated artificially should increase. Only a small proportion of heifers has been inseminated previously because of the inconvenience of heat (estrus) checking and handling for insemination under the typical heifer housing conditions. However, being able to synchronize and inseminate groups of heifers at a preset time should increase the proportion of these animals entering the national dairy herd through AI. This group potentially contains individual animals of the highest genetic merit because the genetic merit of bulls in the bull studs has been increasing. Despite the fact that these heifers have not been selected on their own performance, it is a sound practice to breed them artificially. This should add approximately 1,000,000 animals to the population of dairy animals inseminated artificially by 1990.

B. Beef Cattle

Beef cattle changes and the current inventory are summarized in Table VIII. Cattle numbers have been declining since 1975. However, in 1979 there was a 34% reduction in cow slaughter and a total reduction of 17% in cow and calf slaughter. Consequently, there were more cows on farms in 1980 than there were in 1979. Because of a good calf crop in 1980, the inventory increased to about 116,000,000 by January 1, 1981.

If the usual 10–12-year cycle in beef prevails there should be an upward trend associated with an increase in beef prices. However, with the more competitive prices for fish, chicken, and pork, beef prices have not advanced as in the past. Also, the amount of beef available is not wholly dependent upon cattle numbers but also upon market weight. In 1975, more cattle were slaughtered than in 1978, yet the pounds of beef marketed in 1978 exceeded 1975 by about 4%. A higher proportion of the cattle (77%) came from the feed lots in 1979, whereas only about 50% of the cattle were put through feedlots and fed concentrates in 1975. Political decisions also affect supply and demand.

It should be pointed out (Reid and White, 1978) that edible protein can be made by beef cattle on all-forage diets, requiring little fossil fuel. Animals on pasture harvest most of their energy for growth by consuming grasses, thereby utilizing energy from the sun through photosynthesis. In an earlier section, it was pointed out that dairy cattle might produce up to 5400 kg of milk per year on such a system, conserving concentrates and fossil fuels. Beef cattle would require about 22–23 months to reach market weight of 500 kg under such a system. With a

TABLE VIII

Beef Cattle of All Types on Farms

Year	Inventory (Jan. 1)	Commercial slaughter
1955	96,592,000	39,452,000
1960	96,236,000	36,644,000
1965	109,000,000	40,959,000
1970	112,369,000	39,557,000
1975	132,028,000	46,870,000
1976	127,976,000	48,700,000
1977	122,810,000	48,080,000
1978	116,375,000	44,272,000
1979	110,864,000	36,805,000[a]
1980	113,000,000[a]	

[a] Estimated. These values are based upon USDA reports on Meat Animals and Livestock Slaughter and predictions by the Department of Agricultural Economics, Cornell University (1980).

TABLE IX

Current Statistics on Frozen Beef Bull Semen Produced in the United States

Item	Number	Increase from previous year (%)
Beef cows, Jan. 1, 1980	36,983,000	—
Domestic sales of beef semen, 1979	1,086,339	6.5
Export sales of beef semen, 1979	240,032	60.7
Custom freezing of beef semen, 1979	1,124,586	10.2

typical high concentrate feeding system beef cattle would reach market weight of 500 kg aobut 10 months sooner.

Beef cattle numbers should remain rather stable during the next decade. Any decrease in per capita beef consumption may be balanced by less concentrate feeding, thus requiring relatively more cattle to supply the required pounds of beef per year. The proportion inseminated artificially will continue to be small.

Table IX gives recent statistics on breeding units of beef semen produced in the United States. From these statistics, it can be seen that the largest increase on a percentage basis was in the export market.

Custom freezing is likely to continue at a high level. There are many breeds of beef cattle and varying objectives of beef cattlemen. Many prefer to use semen from their own bulls. With the high interest rates, most will find it difficult to borrow money. If custom freezing of semen is cheaper than buying breeding units from national AI organizations, the present financial climate may tend to increase the number of breeding units produced by custom freezing operations.

Many of the breed organizations are sponsoring breed testing programs that will produce performance- and progeny-tested bulls. The AI industry likely will purchase some of these bulls. Many breeders interested in genetic gains and many commercial cattlemen interested in better calves for market will avail themselves of the opportunity to purchase this semen.

Rotational crossbreeding is a sound system for producing vigorous calves to raise for the beef market. Artificial breeding is the ideal solution for such a program, as cattlemen need only maintain a line of females that can be crossed with semen from bulls of a different breed.

1. Cost of Semen

Semen costs are similar to those discussed under dairy cattle, but one difference is that most beef cows are inseminated by persons working directly for the cattle unit. Insemination costs also should be comparable, provided there are facilities for handling the cattle through corrals and chutes. If these must be constructed, this can easily add $20.00 to $50.00 per cow initially depending upon herd size. Detection of estrus could add a cost of $5.00 per cow.

2. Semen Processing and Banking Semen

The value of banking semen for beef bulls will be less than for dairy bulls because beef bulls can be performance-tested. Also, progeny testing requires less time than for dairy bulls.

3. Sexing Semen

This could have a dramatic effect on the beef industry. Foote and Miller (1971) projected that 100% sexing control could have an annual potential benefit of $200,000,000. This was based upon the replacement of 10,000,000 female calves with male calves produced through sexing of semen. At the time of the prediction in 1970, the market value for steers was about $20.00 more than for heifers. Steers are heavier at weaning and gain weight more efficiently. Now the margin is much greater than in 1970. This potential method of biological control is advantageous over the use of injected or oral additives because of the potential hazards of some additives such as diethylstilbestrol.

4. Estrous Cycle Regulation and Embryo Transfer

Estrous cycle regulation is discussed by Inskeep and Peters (Chapter 13, this volume). This technique has the potential of increasing the number of beef cows inseminated by several million, advancing the calving date, and increasing the quality of calves produced. Such a program could be worth $100,000,000 annually, but the costs of the program could offset this potential under various conditions (see Chapter 13, this volume). Embryo transfer will be a modest program in beef cattle, unless twinning is feasible (see Chapter 3, this volume).

C. Other Species

The opportunities are greatest in cattle but the principles apply to all species. For example, fertility likely would be improved in horses, if selection of the more fertile males had been practiced for several generations, as has occurred in AI of dairy cattle. The benefits of research on information obtained through AI programs also would accrue benefits.

Many of the successful techniques of AI used for domestic animals are being modified successfully with exotic and/or endangered species (Watson, 1978; Seager *et al.*, 1980). Aquatic animals will be an ever increasing source of animal protein in the future (see Chapter 5, this volume). Successful research on the preservation of spermatozoa by freezing in several aquatic species has been patterned after the research with farm animals (Zell *et al.*, 1979; Erdahl and Graham, 1980). This also is true of human AI (Sherman, 1978).

D. Investments in Research: Needs, Costs, and Benefits

The yields from research in AI have been enormous. This must continue if we are to move forward or even prevent regression. Some examples of needed genetic research areas associated with AI are studies of sire effects on (1) protein production in milk, (2) resistance to mastitis and other disease, and (3) dystocia and calving problems. In the area of fertility, studies are needed on predicting sire's fertility from semen quality (Oltenacu *et al.*, 1980) and isolating various problems of sire and maternal effects on fertilization and embryo mortality. Research on sex control with spermatozoa would require extensive AI programs for testing and application. New opportunities now exist for improved utilization of knowledge obtained by modeling and by systems analysis techniques (Rounsaville *et al.*, 1979).

Another potentially extremely important area for study is the relationship between testicular function of males and reproductive development and performance in females (Zimmerman, 1979). Recently it has been found (Coulter and Foote, 1979) that testis size is highly heritable in both dairy and beef bulls ($h^2 = 0.67$ to 0.69). Testis size is important, per se, in AI because of its high correlation with sperm production ($r = 0.8$ to 0.9). Preliminary evidence points to a relationship between testis size and ovulation rate in related female mice and sheep. Also age at puberty of beef heifers is genetically correlated with testis size of related sires.

1. Research: The Endangered Species?

Historically (and in this Chapter), agricultural research has been cited repeatedly as a prime example of research that has reaped benefits. The savings in milk production alone have been worth billions of dollars. It is estimated that during the next 10 years the genes transmitted through AI organizations will be worth a total of about $800,000,000 more than if dairymen use their own bulls. This may be an underestimate because present sire selection programs are superior to previous ones.

Gerrits *et al.* (1979) summarized a national study of research and benefits in reproductive physiology. The figures are rounded off here.

Total agricultural research	Dollar amount (millions)
Effort in reproduction	23
Potential savings, given research support to solve problems	
Dairy cattle	300
Beef cattle	800
Sheep	10
Swine	500
Poultry	85
Total	$1695

38   Robert H. Foote

Even if these figures should be somewhat inflated, the need is evident. The dollars put into research in the dairy cattle AI program can serve as a model. The researchers years ago identified the needs and proceeded to tackle the problems with modest budgets. Today research funds in agriculture are minimal. Much of the money that is available is earmarked for predetermined highly specified areas of research. The creativity and success of the free enterprise system of research has been seriously eroded by the daily dictums of modern bureaucracy.

REFERENCES

Bartlett, D. E. (1980). *Proc. 9th Int. Congr. Anim. Reprod. Artif. Insemin., Madrid* **2,** 271.
Berndston, W. E., and Pickett, B. W. (1978). *In* "The Integrity of Frozen Spermatozoa" (A. P. Rinfret and J. C. Petricciani, eds.), pp. 53–77. Natl. Acad. Sci., Washington, D.C.
Coulter, G. H., and Foote, R. H. (1974). *Proc. 5th Tech. Conf. Anim. Reprod. Artif. Insemin.,* Columbia, Missouri, pp. 67–72.
Coulter, G. H., and Foote, R. H. (1979). *Theriogenology* **11,** 297.
Department of Agricultural Economics (1980). Dept. Agricultural Economics, 137 pp. Cornell University, Ithaca, New York.
Erdahl, D. A., and Graham, E. F. (1980). *Proc. 9th Int. Congr. Anim. Reprod. Artif. Insemin., Madrid.* **2,** 317.
Everett, R. W. (1975). *J. Dairy Sci.* **58,** 1717.
Foote, R. H. (1975). *Theriogenology* **3,** 219.
Foote, R. H. (1978). *In* "The Integrity of Frozen Spermatozoa" (A. P. Rinfret and J. C. Petricciani, eds.), pp. 144–162. Natl. Acad. Sci., Washington, D.C.
Foote, R. H. (1980). *In* "Reproduction in Farm Animals" (E. S. E. Hafez, ed.), 4th ed., pp. 521–545. Lea & Febiger, New York.
Foote, R. H., and Miller, P. (1971). *In* "Sex Ratio at Birth—Prospects for Control" (C. A. Kiddy and H. D. Hafs, eds.), pp. 1–10. Symp. Am. Soc. Animal Sci., Albany, New York.
Foote, R. H., Henderson, C. R., and Bratton, R. W. (1956). *Proc. 3rd. Int. Cong. Anim. Reprod., Cambridge.* **3,** 49.
Foote, R. H., Oltenacu, E. A. B., Kummerfeld, H. L., Smith, R. D., Riek, P. M., and Braun, K. (1979). *Br. Vet. J.* **135,** 550.
Gerrits, R. J., Blosser, T. H., Purchase, H. G., Terrill, C. E., and Warwick, E. J. (1979). *In* "Animal Reproduction" (H. W. Hawk, ed.), pp. 413–421, Allenheld, Osmun, and Co., Montclair, New Jersey.
Graham, E. F. (1978). *In* "The Integrity of Frozen Spermatozoa" (A. P. Rinfret and J. C. Petricciani, eds.), pp. 4–44. Natl. Acad. Sci., Washington, D.C.
Gustafsson, B. K. (1978). *Cryobiology* **15,** 358.
Henderson, C. R. (1973). In "Proc. of Animal Breeding and Genetics Symp. in Honor of J. L. Lush", pp. 10–41. Amer. Soc. of Anim. Sci., Champaign, Illinois.
Iritani, A. (1980). *Proc. 9th Int. Congr. Anim. Reprod. Artif. Insemin., Madrid* **1,** 115.
Lake, P. E. (1980). *Proc. 9th Int. Congr. Anim. Reprod. Artif. Insemin., Madrid* **2,** 501.
Larsson, K. (1978). *Cryobiology* **15,** 352.
Oltenacu, E. A. B., Foote, R. H., and Bean, B. (1980). *J. Dairy Sci.* **63,** 1351.
Perry, E. J., ed. (1968). "The Artificial Insemination of Farm Animals," 4th ed. Rutgers Univ. Press, New Brunswick, New Jersey.
Pickett, B. W., and Berndtson, W. E. (1974). *J. Dairy Sci.* **57,** 1287.

Pickett, B. W., Berndtson, W. E., and Rugg, C. D. (1976). *Proc. 10th NAAB Beef Artif. Insemin. Conf.* pp. 54–77.

Pimentel, D., Oltenacu, P. A., Nesheim, M. C., Krummel, J., Allen, M. S., and Chick, S. (1980). *Science (Washington, D. C.)* **207,** 843.

Pursel, V. G. (1979). *In* "Animal Reproduction" (H. W. Hawk, ed.), pp. 145–157. Allenheld, Osmun, and Co., Montclair, New Jersey.

Reid, J. T. (1978). *In* "Agricultural and Food Chemistry: Past, Present, Future" (R. Teranishi, ed.), pp. 120–135. AVI Press, Westport, Connecticut.

Reid, J. T., and White, O. D. (1978). *In* "New Protein Foods" (A. M. Altschul and H. L. Wilcke, eds.), Vol. 3, pp. 116–143. Academic Press, New York.

Rounsaville, T. R., Oltenacu, P. A., Milligan, R. A., and Foote, R. H. (1979). *J. Dairy Sci.* **62,** 1435.

Rowlands, I. W., and Allen, W. R., eds. (1979). Equine Reproduction II. *J. Reprod. Fertil.,* Cambridge, Eng.

Saacke, R. G. (1978). *Proc. 7th Tech. Conf. Anim. Reprod. Artif. Insemin.,* Columbia, Missouri, pp. 3–9.

Salisbury, G. W., and Lodge, J. R. (1978). *In* "The Integrity of Frozen Spermatozoa" (A. P. Rinfret and J. C. Petricciani, eds.), pp. 92–125. Natl. Acad. Sci., Washington, D.C.

Salisbury, G. W., VanDemark, N. L., and Lodge, J. R. (1978). "Physiology of Reproduction and Artificial Insemination of Cattle," 2nd ed. Freeman, San Francisco, California.

Seager, S. W. J., Wildt, D., and Platz, C. (1980). *Proc. 9th Int. Congr. Anim. Reprod. Artif. Insemin., Madrid* **2,** 571.

Sherman, J. K. (1978). *In* "The Integrity of Frozen Spermatozoa" (A. P. Rinfret and J. C. Petricciani, eds.), pp. 78–91. Natl. Acad. Sci., Washington, D.C.

Sullivan, J. J. (1978). *Cryobiology* **15,** 355.

Van Vleck, L. D., and Everett, R. W. (1976). *J. Dairy Sci.* **59,** 1802.

Watson, P. F., ed. (1978). "Artificial Breeding of Non-Domestic Animals," Symp. Zool. Soc. London, No. 43. Academic Press, New York.

Zell, S. R., Bamford, M. H., and Hindu, H. (1979). *Cryobiology* **16,**

Zimmerman, D. (1979). *In* "Animal Reproduction" (H. W. Hawks, ed.), pp. 131–142. Allenheld, Osmun, and Co., Montclair, New Jersey.

3

The Embryo Transfer Industry

GEORGE E. SEIDEL, JR., AND SARAH M. SEIDEL

I. CHARACTER OF THE INDUSTRY

A. Development

From 1891, when Heape first reported the transfer of a rabbit embryo from the reproductive tract of one female to that of another, to 1971, embryo transfer

41

NEW TECHNOLOGIES IN ANIMAL BREEDING
Copyright © 1981 by Academic Press, Inc.
ISBN 0-12-123450-9

remained a laboratory tool used to study the process of reproduction (Betteridge, 1981). However, when European dual-purpose breeds of cattle (exotics), such as Limousin and Simmental, became the rage in North America, Australia, and New Zealand, cattle interests offered large economic incentives to veterinarians and animal scientists to develop embryo transfer into a breeding tool (Carmichael, 1980; Schultz, 1980). They wanted to rapidly amplify the reproduction of purebred exotic females, which were scarce because of international health and trade restrictions. Their primary motive was profit from the sale of a rare and fashionable calf; improvement of the gene pool was a secondary consideration (with a few notable exceptions).

Thus, the technology of embryo transfer in farm animals was developed into a commercially applicable product not with research funds from government, but with money from cattle breeders and speculators. It was demanded as a service by consumers rather than promoted by extension agents among hesitant farmers and ranchers. Demand outstripped the sophistication and maturity of the technology, and the genetic worth of the donors or the effect of the technology on the normality of the offspring was not questioned.

To date, greater than 90% of commercial embryo transfer activity has been with cattle, and, accordingly, nearly all of the material in this chapter concerns work with cattle. Nevertheless, the potential of the technique has also captured the enthusiasm and imagination of breeders of other farm animal species and conservationists concerned with the preservation of endangered species. To some extent, effort and resources have been invested before the limitations and appropriate uses of embryo transfer have been found.

A depressed cattle market in 1977, a glut of exotic progeny, and improved technology resulted in reassessment of the application of embryo transfer in cattle breeding and the growth of an established industry on a more curtailed, but sounder, economic footing. The balance sheet of the industry until this point showed enormous profit for some clients and serious financial losses for others, greatly increased production for some donors of embryos and substantially diminished production for others.

The history of embryo transfer has been the reverse of most new animal breeding technologies in that demand by users rather than promotion by scientists and educators has been the major stimulus for its growth. This growth has taken place in a wide arena, appealing not only to breeders, but to speculators, hobbyists, and journalists as well (Caras, 1980). This technology, like many others, has become a useful, economically justifiable tool for animal breeders, and the commercial embryo transfer industry grows steadily. In 1979, in excess of 17,000 pregnancies were recorded in North America as a result of transfer of bovine embryos (Seidel, 1981a). Breeders paid some $20,000,000 for this service. Continued growth of the industry seems assured, since embryo transfer is a key step in many of the developing technologies, such as prenatal sex selection.

B. Groups Who Benefit

Embryo transfer offers breeders of farm animals unique possibilities that will ultimately result in more readily available food and by-products of higher quality. As with most of the new animal breeding technologies, embryo transfer helps to increase efficiency so that the same quantities of animal products can be produced with less feed, energy, and land. Although its impact is much less than that of artificial insemination with respect to genetic improvement and multiplication of that improvement over the population, certain applications of embryo transfer can result in dramatic savings of time, enabling in a single generation what it takes many years to achieve with conventional breeding programs. Thus, animal breeders and consumers both of animal products and of the conserved resources are the largest groups who benefit from embryo transfer.

Embryo transfer is used by both breeders who depend on animal breeding for the majority of their income and those who breed animals as a hobby, an investment, a speculative venture, or for tax benefits. Embryo transfer is costly and involves considerable financial risk. For these reasons, its primary application at present is the production of breeding stock for later promotion and marketing by individuals who have considerable liquid capital. Nevertheless, as the technology improves and costs decline, a growing number of small-scale breeders are using embryo transfer on a limited basis. Moreover, developing countries are able to rapidly upgrade the national herds by importing embryos of the desired breed and pedigree (Bedirian et al., 1979; Nelson, 1981).

Those who provide embryo transfer services, of course, benefit from this technology as do many related industries that play a supportive or parasitic role. Included in this group are companies that purify, test, and market the drugs used, manufacturers of specialized equipment, livestock agents and promoters, and marketing specialists.

Although it is mentioned in last place, the benefit of embryo transfer technology to research is substantial (Seidel, 1979). The technique is pivotal for the study of such diverse subjects as fertilization, metabolism of preimplantation stage embryos, the effects of aging on reproduction, teratogenesis, vertical transmission of viral infections, and cancer. Moreover, the development of the technology as well as its commercial application have generated a pool of data that has greatly increased our understanding of such reproductive processes as the mechanism of ovulation, the phenomenon of superovulation, the incidence of early embryonic death, the pattern of endocrine responses, and natural fail-safe mechanisms, which keep the production of defective organisms to a minimum (Seidel and Seidel, 1979, 1980).

C. Services Offered

Services offered by commercial embryo transfer companies include superovulation, surgical or nonsurgical collection and transfer of embryos in the clinic, nonsurgical collection of embryos on the farm, surgical or nonsurgical transfer of embryos on the farm, cryopreservation of embryos, and determination of the sex of embryos prior to transfer.

There are, generally speaking, two types of embryo transfer companies. One is the clinic, to which the donors are brought for superovulation and collection of embryos. The clinic usually has its own herd of recipients and maintains control over all aspects of the procedures. The second type is the mobile unit, which travels to the donor at her home farm or ranch, and depends to varying degrees on the facilities and assistance of the owner and his staff.

There are advantages and drawbacks to each. The clinic requires a large commitment from the owner in terms of land, capital equipment, staff, and animals. Operating costs are also high, necessitating a steady high volume of business. While the risk of financial loss is greater with the clinic operation because of the greater investment, there is more consistency in the product because successful operation depends on fewer factors outside the operator's control. Clinics are equipped to provide any or all of the services listed above and can accomodate most farm animal species. On-farm collection of embryos, especially for dairy cattle, is often offered as an additional service.

The mobile unit, on the other hand, usually does not offer surgical embryo collection, surgical transfer, cryopreservation, or sex determination services owing to limited facilities. Moreover, the mobile operator is exposed to greater risk of failure due to such errors as incorrect administration of the superovulatory drugs, inaccurate estrus detection, use of an insufficient amount of semen, or faulty insemination because these procedures are sometimes carried out by inexperienced personnel. The mobile operator usually provides services only for cattle and horses (the only species for which nonsurgical procedures have been developed) and for pigs.

The majority of embryo transfer companies offer services primarily for cattle. Embryo collection is carried out nonsurgically almost universally, and at this time, most companies transfer at least some embryos nonsurgically. However, probably less than half of all embryos are transferred nonsurgically. The technique of nonsurgical transfer is developing rapidly, so this statistic will soon be out of date. A small but growing number of companies offer cryopreservation of embryos among their services. Very few advertise sex determination services, and fewer still can do so truthfully.

D. Costs and Success Rates

At present, amounts between $300 and $2000 per pregnancy are charged in direct costs for the transfer of bovine embryos, depending on what services are provided. Costs for horses are higher; costs for sheep, goats, and swine are considerably lower. In addition, there are many indirect costs, such as registration and blood-typing fees, advertising, or loss of production by the donor while she is under treatment. When these costs are added up and adjusted for the postpartum salvage value of the recipient and the current net interest cost of money, it is economically justifiable for a breeder in the United States to use embryo transfer if he can obtain an average of about $2500 per calf at 6 months of age, including both bull and heifer calves. Unfortunately, one sex or the other will frequently be worth much less than the other (Seidel and Seidel, 1978). The direct and indirect costs of embryo transfer are summarized in Table I for four different combinations of services. The calculations are based on the assumptions that the money invested costs 10% and that the donors are healthy, fertile cows worth $10,000. A similar analysis was made by Eller (1981), and a computer model for calculating the exact cost of embryo transfer in any given set of circumstances has been developed by Brem (1979). Some embryo transfer offspring command high prices. For example, the top-selling animal in a recent National Holstein Convention Sale sold for $131,000 (Seidel, 1981a), and the top-selling heifer in the 1980 ABS Americana sale sold for $22,000 (Anonymous, 1980a).

Success rates with embryo transfer are highly variable for most species. The component that contributes most to this variability is the wide range of response to superovulatory gonadotropins, both among and within individuals. A single treatment of a cow can result in anywhere from 0 to 30 or more embryos, with a mean of 6 or 7 normal ones. On the average, one can expect one-half of the total number of pregnancies to be produced by only one-fourth of the donors (Seidel, 1980a). Seidel (1980a) reported that of 64 superovulated donors, 22% had 0 pregnancies (or no embryos were recovered), 25% had 1–2 pregnancies, 25% had 3–4 pregnancies, and 18% had 5–12 pregnancies each. Nearly 10% of donors treated with superovulatory gonadotropins failed to come into estrus.

Other sources of variability include the skill of technicians, choice of techniques of collection and transfer of embryos, degree of synchrony between the stage of the estrous cycle of the donor and that of the recipients, fertility of the donor, quality of sperm and embryos, nutrition and management, and luck. The rate of fertilization is another variable that is superimposed on all the others. When morphologically normal embryos are transferred surgically by skilled personnel, the rate of survival to term is almost always in the 50–70% range for all farm animal species if animals are appropriately synchronized, the optimal site and timing of transfer are chosen, and other conditions are reasonable (Allen, 1977; Betteridge, 1977; Douglas, 1979).

For cattle, sheep, and goats, the average number of pregnancies that can be obtained with superovulation and embryo transfer represents an increase by a factor of 10 of the annual production of offspring; for swine, this factor is about 3. When attempts are made to recover a single ovum from unsuperovulated cows at every cycle (15 periods of estrus detected per year × 60–70% recovery rate ×

TABLE I

Costs per Pregnancy of Bovine Embryo Transfer[a]

	Dairy cow collected on-farm, embryos transferred to dairy heifers (salvage $1000) at embryo transfer center, one superovulation, weaning at 3 days	Beef cow collected at embryo transfer center, embryos transferred to beef recipients (salvage $600) at embryo transfer center, three superovulations over 6 months, weaning at 6 months	Dairy cow collected on-farm, embryos transferred surgically on-farm to farmer's own heifers (12 available) 10% of embryos recovered must be discarded on the average due to insufficient recipients	Beef cows collected on-farm, embryos transferred nonsurgically to rancher's own synchronized recipients. 10 donors superovulated, 1 week, 80 recipients synchronized
Pregnancy rate (%)	60	60	55	45
Av. no. pregnancies	3.5	10	3	25
Direct costs				
Entrance fee	0	50	0	0
Programming fee	30	0	35	0
Setup fee	0	0	0	40
Pregnancy fee *adjusted* for salvage value of recipient	900	1300	600	500
Board for donor	0	35	0	0
Semen	30	30	35	40
Veterinary fee, donor	0	5	0	0
Travel expenses	30	0	35	40
Collection fee	30	0	35	0

Indirect Costs				
Trucking, donor	0	20	0	0
Trucking, recipients	100	100	0	0
Registration fee	50	50	50	50
Blood typing fees	35	35	35	35
Feed and board, recipient @ 1$/day	200	380	290	470
Feed, board, and delayed breeding of unused recipients	0	0	180[b]	165
Feed and board, calves	180	90	180	90
Veterinary fees, recipients	50	50	50	50
Veterinary fees, calves	10	10	10	10
Interest expense	380	300	325	360
Owner's time @ $10/hour	60	30	100	100
Heat checking and synchronization	0	0	60	30
Extra facilities	0	0	25	15
Loss of recipient's own calf/net	0	0	100	100
Phone, postage, advertising, misc.	50	50	50	50
Total costs				
Total cost	2135	2535	2195	2145
10% calf losses	213	253	219	214
Required average value of calves	2348	2788	2414	2359

[a] Assuming five doses of semen, six transferrable embryos per donor, and calves raised to 6 months of age. (Adapted from Seidel, 1980b.)
[b] Assuming an average of 6 heifers were unused and, thus, held open 30 days and 3 heifers used that did not remain pregnant, thus, were held open 90 days.

80% fertilized, normal embryos), the 12-month yields average five calves. In contrast, when donors are superovulated 3 times (an average of 6–7 normal embryos recovered per attempt, resulting in 3 to 4 pregnancies following transfer; Seidel, 1980a) the yields average 9 to 12 calves. Collection of the single embryo at every cycle is also much more costly in terms of labor. Because the mare does not respond satisfactorily to superovulatory treatment, amplification of reproduction must be achieved by attempting to collect and transfer the embryo at every estrus (10 to 12 estrus periods per year $\times$ 75% recovery rate $\times$ 55% pregnancy rate = 4 offspring, a fivefold increase over the average of 0.8 offspring per year).

For a given individual, the number of progeny produced by embryo transfer can be astounding. In cattle, for example, the number of pregnancies per superovulation ranges from 0 to 20 or more, and some cows have had as many as 50 calves a year (Seidel, 1981a). On the other hand, with occasional individuals, repeated attempts to recover and transfer embryos may fail completely, resulting in lower production of offspring than with conventional breeding. The upper limit of this technology, however, has not been established, partly because of the high costs involved in obtaining statistically meaningful information. If fertile donors were superovulated as frequently as possible throughout their reproductive lives, their production might be increased by a factor of 20 or more on the average.

Although the sex ratio, incidence of abortion, and normality of offspring produced by embryo transfer have not been documented conclusively, data obtained from commercial application of this technology do not indicate that these characteristics differ greatly from those among offspring produced by conventional breeding. Moreover, preliminary studies in the bovine show that (1) the sex ratio of calves produced by embryo transfer is normal (51% bulls); (2) the incidence of abnormalities (1 in 174) is not higher than in the general population; (3) the ratio of number of weaned calves to number of pregnancies (160 of 178; 90%) equals that of the best conventionally managed herds; (4) the abortion rate after 90 days of pregnancy is not increased; but (5) the abortion rate before 90 days of pregnancy may be higher than normal in the case of nonsurgical embryo transfer (46.3% embryonic loss from days 21 to 60 for blastocysts cultured *in vitro* for 24 hours before nonsurgical transfer and 29.3% for blastocysts transferred nonsurgically immediately after recovery (Elsden *et al.*, 1980; Renard *et al.*, 1980a; Seidel, 1981b). In the case of surgical transfer, losses of 9% between days 25 and 60 have been reported (Markette *et al.*, 1980). In the case of cows inseminated artificially, Kummerfeld *et al.* (1978) reported losses of 7% between days 24 and 75, and Bulman and Lamming (1979) reported a 12% loss. Thus, the statistics reflecting the power of embryo transfer to amplify reproduction document an increase in normal offspring, and the success of the technology from this standpoint is high.

The current high costs of embryo transfer technology cannot be recovered in terms of increased meat and milk production alone (see Chapter 12, this volume). From the breeder's point of view, the success of embryo transfer is variable, depending not only on the number of pregnancies, but also on market fluctuations, promotional opportunities, management techniques, and the breeder's goals. Nevertheless, breeders' comments (Beauprez, 1980) and the steady increase in use of embryo transfer services attest to the fact that cattle breeders believe that embryo transfer is profitable, both for the improvement of their herds and for sales of breeding stock.

The volume of commercial embryo transfer has not been documented for species other than cattle, but it is much smaller than for cattle. A preliminary analysis of an uncompleted survey of members of the International Embryo Transfer Society shows that of 97 offering embryo transfer services in the United States as part of a private veterinary practice or as a commercial firm, 93 worked with cattle, 16 with horses, 7 with sheep or goats, 6 with laboratory, small domestic or exotic species, and 3 with pigs.

II. TECHNOLOGY OF EMBRYO TRANSFER

A. Support Technologies

Embryo transfer is a composite technology that requires expertise in many areas. The term *embryo transfer* has become the appellation for superovulation, embryo recovery, short-term *in vitro* culture of embryos, embryo transfer, and sometimes cryopreservation of embryos (Fig. 1). The list is expanding to include diagnosis or predetermination of the sex of embryos and induced multiple births. If newer technologies, such as cloning, were to become technologically and economically viable, the whole embryo transfer enterprise would probably be subsumed under the appellation of "cloning." Reproduction by embryo transfer depends on the support of a host of other technologies, just as conventional reproduction does. Indeed, high quality in support areas is absolutely critical to success. These technologies include estrus detection, estrus synchronization, herd management, herd health, artificial insemination, blood-typing, and elective abortion.

Embryo transfer lore is full of accounts of error in support areas, especially in herd health and the record-keeping aspects of herd management, which have resulted in a full range of misfortune from acute chagrin to professional and financial ruin. For example, failure to detect estrus accurately in recipients or mistimed estrus synchronization results in the transfer of embryos to uteri that are out of synchrony with the uterus of the donor, which can reduce pregnancy rates significantly (Seidel, 1981a). A more serious problem might follow a failure to

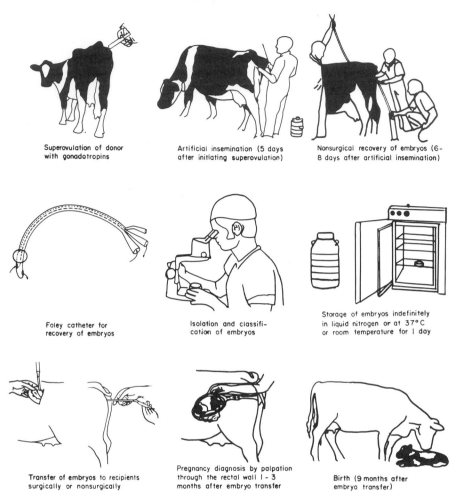

Fig. 1. Synopsis of bovine embryo transfer procedures. (From Seidel, 1981a, copyright 1981, Am. Assoc. Adv. Sci.)

maintain effective quarantine and herd health procedures. With the introduction of brucellosis into a herd of donors at a bovine embryo transfer unit, for example, the compulsory slaughter of these valuable cows (easily worth $10,000 each) could be devastating.

These examples illustrate the often overlooked, but cardinal concept of embryo transfer technology as a whole. While each step is simple and usually a matter of common sense, it must be carried out properly and in coordination with all preceding and subsequent steps, or failure will result.

B. Methods

1. Superovulation

Most embryo transfer donors are treated with pregnant mare's serum gonadotropin (PMSG) or follicle stimulating hormone (FSH) to induce the maturation and ovulation of a larger than normal number of oocytes. Although it is thought that these drugs reverse the normal atresia of maturing follicles and oocytes not destined to ovulate under normal circumstances, their exact mechanisms of action are unknown. The treatments were developed empirically, and adequate procedures are presently available for superovulation of laboratory and farm animal species with the exception of the horse. Despite their having in effect been rescued at the eleventh hour, superovulated ova result in normal offspring with success rates following embryo transfer similar to those achieved with normally ovulated ova (Elsden *et al.*, 1978). Betteridge (1977) and Sugie *et al.* (1980) have summarized techniques for superovulation of farm animals. Superovulation increases the number of normal ova recovered by about a factor of 10 for cattle, sheep, and goats, but only a factor of 2 to 3 for swine.

Treatment regimens vary according to species. The induction of follicular growth by PMSG in the goat, for example, is often combined with treatment with human chorionic gonadotropin (hCG) to induce ovulation (Betteridge and Moore, 1977).

Because timing of treatment is based on anticipated changes in endogenous hormone levels that are not signaled overtly, superovulation should not be induced until at least one normal estrous cycle has been observed. Recent approval by the United States Food and Drug Administration (FDA) of prostaglandin $F_{2\alpha}$ has taken a good deal of the guesswork out of timing treatments. This drug, usually given after the gonadotropins, lyses the corpus luteum, enabling one to begin treatment at any time during the midluteal phase of the estrous cycle in cows, sheep, and goats (Sugie *et al.*, 1980). Unfortunately, it is ineffective in pigs (Anderson, 1980).

At certain times, PSMG and FSH are difficult to obtain from commercial sources. Moreover, both preparations vary considerably from batch to batch in both potency and contamination with other hormones. The FDA has recently declared PMSG a new drug; lawful use of this agent in the United States requires compliance with FDA regulations.

Ovarian response to superovulatory treatment can result in endocrine levels greatly altered from the normal. This may modify subsequent estrous cycles; therefore, an interval of one or two complete estrous cycles between treatments is recommended (Lubbadeh *et al.*, 1980). Cows have been treated 4 or more times without an intervening pregnancy, and most continue to respond to the gonadotropins (Christie *et al.*, 1979), but there are insufficient data to judge the long-term effects of repeating superovulation more than 3 or 4 times. Seidel (1981a) reported no difference in the average number of pregnancies from first, second, and third superovulations of cows. Leidl *et al.* (1980) found that, in the cow, antibody formation induced by injection of therapeutic doses of hCG (occasionally used in recipients on the day of transfer for its luteotropic effect) was not of any practical importance, but similar studies have not been carried out for superovulatory drugs.

Breeders sometimes invest considerable portions of the reproductive life of their donors in superovulation. In a recent study, three successive superovulatory treatments of cows required about 7 months (Seidel, 1981a). If the results are poor, the breeder has lost up to 1 year's production from his valuable cow as well as the fees he paid for the service. That the degree of success cannot be predicted accurately for an individual animal is the greatest drawback of superovulation technology. Not only failure, but also too many pregnancies may sometimes present an economic risk to the breeder.

It may safely be predicted that methods for superovulation will improve with respect to consistency of results. Additional understanding of basic physiological mechanisms will facilitate such efforts. Hormonal parameters will be linked to desired responses and the ability to produce normal embryos should follow quite rapidly. The development of procedures, especially for the horse, will be dictated to some extent by economics and demand.

Exciting new work in superovulatory technology involves the active immunization against androstenedione or testosterone to prevent atresia and increase the frequency of multiple ovulations reliably (Scaramuzzi *et al.*, 1977; Van Look *et al.*, 1978; Scaramuzzi, 1979; Scaramuzzi *et al.*, 1981). The technique has definite commercial potential for sheep and cattle husbandry, and much current effort is directed toward developing and testing a commercial procedure.

2. Embryo Recovery

Unfertilized oocytes, for specialized applications like *in vitro* fertilization, must be collected close to the time of ovulation, either from follicles, the surface of the ovary, or the oviduct. For most applications, embryos are collected sometime between fertilization and implantation, usually after migration to the uterus. In cattle and horses, embryos for commercial purposes are usually recovered 6–9 days after estrus (Elsden *et al.*, 1976; Douglas, 1979). Prior to this time, nonsurgical recovery is not efficacious. After 9 days, recovery and pregnancy rates are slightly reduced, at least with surgical transfer of bovine embryos (Betteridge *et al.*, 1980). For biopsy to determine the sex of bovine embryos, elongating blastocysts 12–15 days after estrus are preferred (Hare and Betteridge, 1978; see Chapter 6, this volume).

Success of embryo recovery is determined not only by the age of the embryos, but also by the technique, the skill of the technician, and other ungovernable and unpredictable factors. For example, in superovulated donors, there are frequently large numbers of follicles that never ovulate and possibly some that develop as if they had ovulated, but never release the oocyte. Further, if the ovaries are greatly enlarged in response to superovulatory treatment or if there is a large number of ovulations, the fimbriae apparently do not gather all of the oocytes into the oviducts, and they are probably lost into the abdominal cavity.

Radically altered endocrine levels as a result of superovulatory treatment may cause the uterus to present an environment that is unfavorable to the development of embryos, resulting in their degeneration. Some evidence indicates that the incidence of degenerate embryos may increase between days 7 and 8 after estrus (when most embryos are collected for transfer). Similar mechanisms may be responsible for the trend of recovery rates to decrease as fertilization rates decrease, as frequently occurs when there is a large number of ovulations (Betteridge, 1977).

Reports of recovery rates are innately inaccurate due to the difficulty of assessing the number of ovulations in superovulated donors. Even by examining the ovaries at surgery, the large number of ovulation sites and the frequently enlarged, abnormal appearance of the ovaries sometimes prevent accurate counts. If nonsurgical recovery methods are used, the number of ovulations can only be estimated by palpation per rectum (Betteridge, 1977).

Preimplantation embryos can be flushed from the oviducts or uteri of excised tracts obtained at hysterectomy or slaughter. This method results in a somewhat higher rate of recovery than other approaches (Sugie *et al.*, 1980), but obviously cannot be repeated on an individual donor.

Surgical recovery is efficient in all species and is the method of choice for sheep, goats, and pigs. It is the only practical means of obtaining embryos that are located in the oviduct. Techniques vary slightly with species (Betteridge, 1977; Sugie *et al.*, 1980; Brackett *et al.*, 1980). It is difficult to pass a collection device through the cervix into the uterus in the sheep, the pig, and the goat. Although this anatomical barrier is not insurmountable, the economic incentive has been insufficient to make the development of nonsurgical recovery techniques imminent, although interest in dairy goats is increasing. Embryos representing 50–80% of the ovulations can be recovered nonsurgically in cattle (Betteridge, 1977). The rate of recovery of the single embryo in horses is reported to be 40–90% (Betteridge, 1977; Imel *et al.*, 1981).

In cows and horses, nonsurgical collection can be repeated an unlimited number of times on an individual donor without reducing her subsequent fertility (Elsden *et al.*, 1976; Imel *et al.*, 1980; Tischner and Bielanski, 1980). Moreover, the requirements in equipment, personnel, and time are minimal so that embryos can be collected at the home farm or ranch of the donor animal. This advantage is offset in the horse by the fact that registration of foals resulting from embryos collected on the farm and transported to the embryo transfer clinic for transfer is not permitted by some breed associations. Nevertheless, it is a significant advantage in the case of dairy cows, since the donor's milk production is not lost, as it would be if she had to be removed to the embryo transfer center. It is sometimes economically feasible to collect a single embryo from a cow following an untreated estrus, between superovulatory treatments, at the first estrus after parturition, or from a young heifer before her estrous cycles have become sufficiently regular to permit reliable superovulation. Currently, surgical recovery is practiced only in cows and mares with suspected pathology of the upper reproductive tract (Bowen *et al.*, 1978), or if ova must be recovered for research purposes at an early stage of development before they migrate to the uterus.

Methods of collecting embryos have not changed appreciably since about 1976, nor are significant advances predicted for the future. Average rates of recovery for farm animal species, based on counts of ovulations, are in the 50–90% range (Betteridge, 1977; Rowe *et al.*, 1980a). It is likely that practically all of the embryos in the reproductive tract are recovered, at least with surgical methods.

3. Estrus Synchronization

For artificial insemination programs and for embryo transfer as well, effective control of estrus is essential. Usually, the objective is to have large numbers of females in estrus at the same time, hence the term estrus synchronization. Effective methods of estrus synchronization have been available for more than 2 decades, but fertility at the synchronized estrus was greatly reduced for most methods. Over the last decade, however, several schemes have become available that result in normal fertility (Guthrie and Polge, 1976; Trounson *et al.*, 1976; Gordon, 1977; Holtan *et al.*, 1977; Hansel and Beal, 1979).

Definitive, controlled comparisons of pregnancy rates with embryo transfer following natural versus induced estrus of recipients have not been made. Available data suggest, however, that there is little difference in pregnancy rates whether recipients are in natural estrus in synchrony with the donor or are treated with prostaglandin $F_{2\alpha}$ or progestogens for synchronization (Betteridge, 1977). Wright (1981) reported a pregnancy rate from nonsurgical embryo transfer of 59% ($n = 1784$) when estrous synchronization was induced among recipients using prostaglandin $F_{2\alpha}$ compared to 58% ($n = 661$) when recipients were in natural estrus synchrony with the donor. Maintenance of a large enough herd to provide a sufficient number of suitable recipients in natural estrus synchrony with a given donor is an enormous expense. If the volume of transfers is high, however, there is no particular economic advantage to inducing synchrony. No matter what the method of synchronization, good estrus detection, record keeping, and herd management are crucial. These aspects represent major obstacles to transfers in the field. Provision of suitable recipients is the greatest single cost in embryo transfer and accounts in large measure for the great difference in cost between this technology and artificial insemination.

The importance ascribed to the technology of estrus synchronization is best illustrated by the fact that more research effort, in terms of man-years or dollars, has been devoted to this problem by universities, government agencies, and private industry over the last decade than any other technology related to reproduction.

The value of being able to breed an animal at a specific time after observing a behavioral response cannot be overemphasized. However, there are additional benefits. Some treatments used in estrus synchronization are also efficacious for inducing puberty and terminating lactational anestrus. Others are useful for treating cystic ovarian disease and certain uterine infections. For scientists, such treatments permit much easier and less expensive experiments.

4. Short-Term Storage of Embryos

Many procedures such as embryo transfer, *in vitro* fertilization, sex determination, and cloning depend heavily on the ability to maintain the viability of embryos for several hours to several days outside of the reproductive tract. Usually, the embryos are cultured *in vitro*, but they may also be stored in the oviduct of a temporary host of another species like the rabbit (Seidel, 1981a). For many applications, the storage system must not only maintain the viability of the embryo, but must also support continued development. There are instances, however, when it is desirable to retard growth to a degree approaching suspended animation so that development can be synchronized with later events. For example, it may be necessary to store embryos until suitable recipients become available for transfer.

Most media and culture systems are adequate for maintaining the viability of the embryo between recovery from the reproductive tract of the donor and transfer to a recipient, normally an interval of from 20 min to more than 24 hr. If embryos must be transported any distance, however, or if suitable recipients are not available, a longer-term storage system is needed, like the oviduct of the rabbit. Embryos from cattle, sheep, and goats tolerate cooling to 0°–10°C for several days (up to 10 days in the case of sheep) without substantial loss of viability (Betteridge, 1977). Pig embryos do not survive cooling below 15°C (Polge, 1977). The physical and nutritional requirements of embryos from farm animal species have not been defined, although adequate culture systems for short intervals have been developed by trial and error (Betteridge, 1977; Seidel, 1979; Sugie *et al.*, 1980). In culture experiments, most embryos have undergone one cleavage division and, in some cases, two or more. Sheep embryos in one trial developed from the 1-cell to the hatched-blastocyst stage *in vitro* (Wright *et al.*, 1976b).

Important parameters of culture systems include temperature, pH, osmolality, light, oxygen tension, and composition of the medium. High oxygen tension is detrimental to embryos (Seidel, 1979; Harlow and Quinn, 1979). Depending on the buffer system of the culture medium, air, 5% CO_2 in air, or 5% CO_2, 5% O_2, 90% N_2 are used for the gas phase. An atmosphere of 5° CO_2, 5% O_2, and 90% N_2 was reported to support the greatest number of cell divisions when 1- to 8-cell cow and sheep embryos were cultured (Wright *et al.*, 1976a,b), but it is awkward to maintain any gas phase other than air since the container must be opened repeatedly to remove embryos for transfer. For this reason, phosphate-buffered media are in general use, even though these media are not optimal for embryos.

Many media, such as Ham's F-10, TCM-199, and Dulbecco's phosphate-buffered saline, have been used for embryos (Betteridge, 1977). They contain various concentrations of inorganic salts, carbohydrates, amino acids, vitamins, and nucleic acid precursors (Sugie *et al.*, 1980). Some are exceedingly complex, others very simple. Nearly all contain either heat-inactivated serum (usually 5–20%) or bovine serum albumin (0.1–3.2%) (Sugie *et al.*, 1980). Untreated serum is toxic. Embryos are very sensitive to bacterial contamination; therefore, most media contain antibiotics and/or are filtered through a 0.45-μm filter (Betteridge, 1977). Glass-distilled, deionized water should be used to eliminate heavy metals and other contaminants.

For short-term storage, it is preferable to use simple media like phosphate-buffered saline or commercially available media with a buffer like HEPES, requiring no controlled atmosphere. Temperatures between 15° and 37°C, osmolalities between 270 and 310 mOsM/kg, and pH of 7.0–7.8 provide suitable conditions for embryo survival in a variety of balanced salt solutions (Seidel, 1977; Seidel, 1979; Sugie *et al.*, 1980). The use of incandescent light with interference filters is recommended for examining embryos, and the time under the microscope should be kept to a minimum (Hirao and Yanagimachi, 1978).

The metabolism of preimplantation embryos is being studied intensively in laboratory species and to an increasing extent in farm animal species. Culture systems are being developed that will support normal development of immature oocytes through meiosis, fertilization, and cleavage to the hatched-blastocyst stage. These systems will increase the flexibility of embryo transfer techniques and enable new applications, but will probably not improve the success rates of routine recovery and transfer. They will be of much more significance to techniques for *in vitro* maturation of gametes, *in vitro* fertilization, sex determination, cloning, and other forms of genetic engineering that involve prolonged manipulation of gametes and embryos outside of the reproductive tract.

5. Deep-Freezing of Embryos

Freezing embryos in liquid nitrogen for indefinite storage is covered in detail by Leibo (see chapter 7, this volume). Clearly, if embryos could be frozen with little loss in viability and if those procedures were simple and inexpensive, the character of commercial embryo transfer would change dramatically. There would be much greater flexibility in timing work, and fewer recipients would need to be maintained. Even though the survival rate of frozen-thawed bovine embryos is only about 70%, it is already profitable to freeze embryos if insufficient recipients are available and in specialized situations such as exporting embryos.

6. Embryo Transfer

Embryos in early stages of development should be deposited in the oviducts; this must be done surgically. Surgery is also required for uterine transfer in sheep, goats, and pigs and is still the predominant method in cows and horses. Foote and Onuma (1970), Gordon (1975), Betteridge (1977), Seidel (1979), and Sugie *et al.* (1980) have reviewed surgical transfer methods. Interactions of a number of factors determine the success of surgical transfer: age and quality of embryos, site of transfer, degree of synchrony between estrus cycles of the donor and recipients, number of embryos transferred, *in vitro* culture conditions, skill of personnel, and management techniques. When these variables are within an acceptable range, the pregnancy rates following transfer of normal bovine embryos are comparable to the first-service pregnancy rates with artificial insemination.

Success rates with embryos transferred to the oviduct (1–3 days after estrus for cows, goats, and sheep, and 1–2 days for swine) are generally lower than rates with older embryos transferred to the uterus (Betteridge, 1977; Moore, 1977). In cows, sheep, and goats, success rates do not differ appreciably between days 5 and 10 after estrus. Most transfers are made between days 5 and 8 in these species because older embryos are more difficult to find and handle (Betteridge, 1977; Elsden *et al.*, 1978; Shea, 1981). Success rates with the transfer of pig embryos drop markedly after 6 days post-estrus (Polge, 1977). The side of the uterus to which embryos are transferred is a critical factor in the cow. Embryos transferred to the uterine horn contralateral to the recent ovulation result in low pregnancy rates (Newcomb *et al.*, 1978; Sreenan, 1978b; Seidel, 1981b). A single uterine horn in the cow frequently does not support more than one pregnancy (Rowson *et al.*, 1971); therefore, if two embryos are transferred, one should be deposited in each horn.

In cattle, the reproductive tract can be exposed for surgical transfer of embryos either through a midline laparotomy with general anesthesia or through a flank incision with local anesthesia. The midline approach may be less traumatic to the reproductive tract and permits better exposure in some animals. The flank approach, on the other hand, is faster, less expensive, and it eliminates the risks inherent with the use of general anesthesia. It can also be done in the field with a minimum of equipment. Karihaloo (1981) reported a 5% lower pregnancy rate using the flank approach, but the savings in time over the midline approach enable a large enough increase in the number of embryos transferred to more than offset this loss.

Depending on whether general or local anesthesia is required and whether a flank or midline incision is used, the surgical transfer of an embryo takes about 10–30 min on the average. If a large number of embryos must be transferred, this represents a considerable cost in time. A bare minimum of two or three highly skilled workers is required, but efficiency is increased if additional help is available. Salaries of the support staff (for herd management, health, record keeping, and administration) comprise an even larger component of costs. Personnel requirements, therefore, account for a large share of the high costs of surgical embryo transfer, which, with the complexity of surgery, limit the applicability of surgical embryo transfer to very valuable embryos.

Pregnancy rates with cattle embryos are very low outside of a range of tolerance of 48 hours either side of exact synchrony between the reproductive cycles of the donor and recipient, although a recent reevaluation of the literature suggests that the degree of asynchrony that can be tolerated routinely is not yet known (Seidel, 1981b). Embryos from sheep and goats may tolerate a slightly greater degree of asynchrony than those of cattle (Seidel, 1979).

Pregnancy rates are highest in cattle when two embryos are transferred. Rates of 75% are common (Betteridge, 1977). However, embryo survival rates in these studies are in the range of 50–60%. Rowson *et al.* (1969), for example, reported a pregnancy rate of 91% compared to an embryo survival rate of 46%, a dramatic difference. The transfer of more than three embryos does not increase pregnancy rates (Betteridge, 1977), nor does it appreciably increase the net number of progeny. The transfer of too many embryos per uterus will, in fact, reduce the pregnancy rate to near zero in cattle.

Embryo quality is an important determinant of the success of transfer. Gross morphological characteristics correlate fairly well with pregnancy rates on a population basis, but they are of limited value in determining the potential of an individual embryo to develop following transfer (Seidel, 1981b). The most accurate predictor of success is appropriate stage of development (Kunkel and Stricklin, 1978). Success is very low with embryos retarded 2 days or more in comparison to the expected stage of development (Elsden *et al.*, 1978).

Other parameters for evaluating embryos include shape, color, number and

---

Fig. 2. Bovine ova. (A) Unfertilized oocyte recovered from the ovary. Follicle cells were removed mechanically. The outside layer is the zona pellucida. (B) Four-cell embryo recovered from the oviduct on day 2 of pregnancy (estrus, day 0). Note perivitelline space between the cells and the zona pellucida. (C) Eight-cell embryo recovered from the oviduct on day 3 of pregnancy. (D) An 8- to 16-cell embryo (morula) recovered from the oviduct on day 4 of pregnancy. (E) Very early blastocyst stage; approximately 60 cells recovered from the uterus on day 7 of pregnancy. The inner cell mass from which the fetus will develop will form to the right. (F) Expanded blastocyst with > 100 cells recovered from the uterus on day 8 of pregnancy. The inner cell mass is to the right. All photomicrographs were taken with differential interference contrast (Nomarski) optics. Magnification approximately × 300. (From Seidel 1981a, copyright 1981, Am. Assoc. Adv. Sci.)

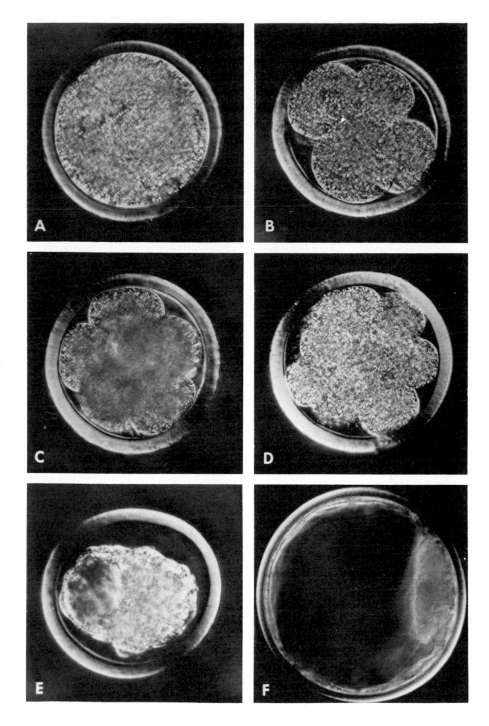

TABLE II

Recent Studies in Which Single Bovine Embryos Were Transferred Nonsurgically

Author	Treatment	N	% Developing[a]	Comments
Greve and Lehn-Jensen (1979)	Days 6–7	32	56	Cows, epidural anesthesia
Halley et al. (1979)	Day 6	170	24	Brahman crosses, large
	Day 7	274	26	technician differences
	Day 8	142	42	
Jillela and Baker (1978)	Days 6–7	6	67	Recipients were pluriparous cows
	Days 8–9	15	47	
Newcomb et al. (1978)	Ipsilateral	20	20	Also surgical transfer contralateral
	Contralateral	20	35	Also surgical transfer ipsilateral
Newcomb (1979)	Exact synchrony	16	62	Heifers, days 7–10
	± 1 Day asynchrony	54	44	
	Exact synchrony	10	80	Cows, day 7
Sreenan and McDonaugh (1979)	Days 9–13	36	47	Ipsilateral, not inseminated
	Days 9–13	25	44	Contralateral to insemination

Reference		n	%	
Christie et al. (1980)	hCG	20	65	Heifers, day 7
	Control	20	50	Heifers, day 6
Rowe et al. (1980b)	—	20	75	Heifers, days 6–8
	Operator 1	40	58	
	Operator 2	40	35	
Renard et al. (1980a)	Immediate transfer	83	49	Heifers, days 10–12
	24-hour culture	45	33	
Renard et al. (1980b)	Glucose uptake	13	61	Heifers, days 10–11
	No uptake	14	14	
Schneider et al. (1980)	Operator 1	153	48	Heifers, days 6–8
	Operator 2	145	53	
	Operator 3	125	28	
Tervit et al. (1980)	Good embryos	22	59	Heifers, days 5–9
	Poor embryos	18	17	
Shea (1981)	Good embryos	679	56	Days 7–9
	Fair embryos	130	44	
Curtis et al. (1981)	Technician 1	60	35	Days 6–8, cows and heifers
	Technician 2	60	13	
Wright (1981)	Cows	2200	59	Days 6–8, Brahman crosses
	Heifers	245	58	

[a] Most data represent early pregnancy diagnosis rather than calving rates.

compactness of cells, size of the perivitelline space, and the number and size of vesicles. An ideal embryo is compact and spherical with cells of similar size and uniform color and texture (Fig. 2). The zona pellucida is even; the perivitelline space is empty and of regular diameter (Seidel et al., 1980). Nevertheless, some embryos that appear abnormal by these standards develop into normal young.

The quality of ova has been shown to be influenced by a number of factors. The viability of ova from prepuberal animals is lower than that of ova from adults (Seidel, 1981b). The uterus of superovulated cows, particularly when the response to superovulation is good, is probably harmful to embryos because of inappropriate hormonal levels. Removing embryos after 3–4 days and storing them for 2–3 days in the rabbit increases the success of transfers (Betteridge, 1977). However, the difference in pregnancy rates between superovulated and nonsuperovulated bovine embryos seems to be small (Elsden et al., 1978).

Culling embryos on the basis of stage of development and appearance increases pregnancy rates markedly. Wright (1981) reported pregnancy rates of 44 and 53%, respectively, with morulae and advanced morulae compared to 65, 66, and 64% ($P < .05$) with early blastocysts, blastocysts, and advanced blastocysts. Similarly, Halley et al. (1979) reported a 14% pregnancy rate for morulae, 22% for blastocysts, and 30% ($P < .01$) for expanded blastocysts from Brahman cattle. A number of similar studies have been summarized (Seidel, 1981b). In commercial situations, however, embryos represent a considerable potential profit and even abnormal ones with scant chance of developing into pregnancies are sometimes transferred. With high pregnancy fees, the percentage of apparently abnormal embryos that do develop is frequently high enough (because of inaccurate identification of poor risks) to pay for the ones that do not. Sometimes, pregnancy rates from experimental transfers are better than those reported by commercial units because questionable embryos are discarded. Research to improve evaluation of embryos, such as the development of fluorescent live–dead staining techniques (Schilling et al., 1979a,b) and to control factors that influence the quality of embryos would be of great value. Considerable time and money could be saved if all embryos with less than a 10% chance of developing into a fetus could be identified accurately and discarded.

Possible reasons for the failure of transfers include (1) loss of the embryo during transfer either because it adhered to the equipment or was not deposited in the lumen of the uterus, (2) excessive trauma to the reproductive tract, which has been shown to reduce pregnancy rates to zero in cattle (Nelson et al., 1975), and (3) infertility of the recipient. Crossbred recipients with hybrid vigor should have slightly higher fertility (Seidel et al., 1980); apart from this, breed is not a significant factor. Age and parity are not critical except in the case of senescent cows. Often, however, the only animals available as recipients are ones culled from breeding herds for failure to become pregnant. Also, recipients are often reused if they did not become pregnant following the first transfer. Recently, pregnancy rates after the first transfer were reported to be 60% compared to 44% after the second transfer (Nelson *et al.*, 1980). When the recipients that had failed to become pregnant after two transfers were bred artificially, however, conception rates were equal to those of the normal population.

Surgical transfers will remain the method of choice for sheep, goats, and pigs in the foreseeable future. For cows and horses, however, nonsurgical methods will probably replace surgical transfer within 1–2 years.

Nonsurgical embryo transfer is an adaptation of the artificial insemination technique. The embryo is introduced into the uterus transcervically through a 0.25-ml plastic straw in a straw gun or similar device (Brand and Drost, 1977; Sreenan, 1978b; Douglas, 1979; Schneider and Hahn, 1979). Recent results with cattle are presented in Table II. Pregnancy rates following the nonsurgical transfer of horse embryos have been reported to be 34% (Douglas, 1979) and 27% (Imel *et al.*, 1981; pregnancy rates of surgical controls in this study were 53%). It is much faster and cheaper to transfer embryos nonsurgically. Requirements for personnel and equipment are similar to those for artificial insemination. Reported success rates, however, are generally much lower than those with surgical transfer. The cause of reduced pregnancy rates is unknown, although it may be some combination of trauma to the reproductive tract during passage of the straw gun (resulting in the release of substances toxic to the embryo), infection of the uterus, improper placement of the embryo, expulsion of the embryo, and other unknown factors (Seidel, 1981b).

Nevertheless, there are numerous embryo transfer units in North America that transfer all embryos nonsurgically. It is very likely that one-half of the commercial transfer pregnancies in cattle in North America in 1981 will be done nonsurgically, even though success rates are only 70–90% of those obtainable with surgical transfer.

C. Applications

Adequate technology and demand are available to make the following applications feasible now.

1. Obtain More Offspring from Valuable Females

In cattle, for example, reproduction of a fertile female using repeated superovulation can be increased from a single calf (twins are born in a few percent of pregnancies) to an average of 10 per year (Seidel, 1981a). If the annual reproduction of elite females were increased 10-fold, the selection pressure on females could be increased by a like factor. In dairy cattle, for example, 70–90% of female calves must be reared as replacements for animals culled for reasons other than low production. If the replacements were selected from the top 10% of the herd by amplifying the reproduction of superior cows by embryo transfer, the rate of genetic gain from the female side could be increased 3 to 4 times (Seidel, 1981a; see Chapter 12, this volume). The National Cooperative Dairy Herd Improvement Program reported that the 275 embryo transfer donors on record were at the 75th percentile for genetic value for milk production, whereas their 50 progeny by embryo transfer that were old enough to have records were at the 87th percentile (Seidel, 1981a). The increased genetic value over the population average that these cows would transmit to offspring was estimated to be 112 kg of milk for the donors and 155 kg for their embryo transfer offspring.

2. Obtain Offspring from Infertile Females

The uterus often loses the ability to support a pregnancy before the ovary ceases to produce ova. When ova from senescent animals are fertilized and transferred to a young uterus, normal offspring result (Maurer and Foote, 1971). Infertility caused by age or disease has often been circumvented in cattle by embryo transfer (Bowen et al., 1978; Elsden et al., 1979). This is an especially important application for horses.

3. Export and Import

Embryos are easier and less expensive to transport than adult animals and success rates can be high (Bedirian et al., 1979). Moreover, adaptation to a new environment is easier when the individual is introduced as an embryo than as an adult because passive immunity to the local pathogens is acquired through the colostrum of the recipient and because the immune system develops in the appropriate environment (Seidel, 1981a). Another advantage of importing embryos is that there is much less chance of introducing new diseases than with importation of animals (Waters, 1981).

4. Test for Mendelian Recessive Traits

It is exceedingly difficult to test whether a monotocous female is a carrier of undesirable traits without embryo transfer technology to amplify her reproduction. Proof that an animal is not a carrier of an undesirable trait rests on a statistical probability test requiring at least eight normal offspring from matings with a homozygous carrier (Seidel, 1981a). For a non-litter-bearing species, it would take almost the entire reproductive lifetime of a female to prove that she is free of a particular recessive allele. Furthermore, animals with homozygous alleles for an undesirable trait are normally destroyed; therefore, very few females are available for testing possible male carriers. With embryo transfer, an animal can be tested within 1 year with only one homozygous mate (Baker et al., 1980; Johnson et al., 1980). To avoid the unwitting widespread transmission of undesirable traits through artificial insemination, tests of this kind are essential.

5. Introduce New Genes into Specific Pathogen-Free Herds

Confinement breeding of large herds is the production trend in the swine industry. Closed herds, and especially specific pathogen-free herds, are potentially compromised when new genetic material is introduced. If new genes are introduced by means of embryo transfer, the risk of disease transmission is greatly reduced (Day, 1979). Similarly, valuable genetic material from quarantined herds can be rescued by embryo transfer if the pathogen in question is known not to infect the embryo (Waters, 1981).

6. Increase the Population Base of Rare or Endangered Breeds or Species

Embryo transfer is effective in this regard if the embryos of the rare species or breed can be transferred to recipients of a more plentiful species or breed, such as from mouflons to domestic sheep or from yaks or American bison to domestic cattle. On a population basis, conventional reproduction is usually more efficient than embryo transfer since the increased reproduction of an embryo transfer donor is achieved at the expense of decreased reproductive rates of the recipients. The reproductive rate of recipients is lowered primarily because they must remain nonpregnant until they receive an embryo. In practice, this means that a recipient may be out of production for 2–6 months. The development of effective embryo transfer techniques for exotic species is felt to be very significant to the preservation of genetic diversity among captive (zoo) endangered species. Durrant and Benirschke (1981) foresee its use to decrease inbreeding depression by (1) exchanging gametes among captive populations of an endangered species, thus increasing breeding choices when costs, quarantine problems, and trauma make shipping of adults impossible, and (2) freezing embryos from one generation for transfer to a later generation. They also envision the increase of populations of endangered species, such as ruffed lemurs, that are of value as models for study of human disease and the use of interspecies embryo transfer for studies on hybridization.

7. Research

Embryo transfer is a *sine qua non* for much research involving *in vitro* studies of gametes or embryos. It is very useful for other research also (Seidel, 1979).

Technology must be improved or developed and economic incentive provided before the following applications are realized.

8. Induce Twinning

The induction of twin births in beef cows would greatly increase the efficiency of feed conversion (Anderson, 1978; Seidel, 1981a). Twinning by embryo transfer will become economical when the cost falls below the average price for newborn calves to be raised for meat ($100 for dairy calves in 1981), and management techniques for expected twin births are refined. Coupled with cryopreservation, sexing, and *in vitro* fertilization, twinning by embryo transfer could significantly increase the production of animal protein by making use of untapped resources such as ovaries from females sent to slaughter and frozen embryos that have been identified as males that are of little value as breeding stock (Seidel, 1981a).

9. Obtain Progeny from Prepuberal Females

This would reduce the generation interval to enable faster genetic progress when selecting for traits that can be recognized prepuberally.

10. Progeny Test Females

At present, it is an advantage to progeny test females only in specialized cases. In inbred herds, for example, genetic progress cannot be made as rapidly as in the general population by selecting males because of very limited numbers. In such situations, it is of relatively more value to apply selection pressure to the females in the herd. For large breeding populations, however, the value of the information obtained by progeny testing females is offset by the length of time it would take to acquire it (Seidel, 1981a). For example, it takes about 5 years to obtain a progeny test for milk production in cattle, and at the end of the test period the average of the population could already be higher than the genetic value that could be contributed by the cows from the current population identified as outstanding by progeny testing.

D. New Technologies Dependent on Embryo Transfer

Many new technologies that will become available to the animal breeder in the coming decades depend on the ability to remove gametes or embryos from the reproductive tract for manipulation and to return them to the reproductive tract for gestation to term. These include sex determination, sex selection, fertilization *in vitro,* production of identical twins, cloning, and the host of technologies grouped under the term genetic engineering. In addition, a number of new industries are proliferating based on embryo transfer: purification of gonadotropins, manufacture of micromanipulators and similar specialized equipment, international marketing of embryos, specialized investment counseling, and certain branches of law (e.g., laws governing the rights of inheritance of an embryo transferred to a woman who is not the genetic mother and is not married to the genetic father.)

III. FACTORS AFFECTING GROWTH OF THE INDUSTRY

A. Trends in Economy of Animal Agriculture

The extent to which embryo transfer technology is used is very sensitive to the profits and losses of the livestock industry as a whole. With cattle in North

America, applications to date have not been closely linked to increased meat or milk production, but rather to the value of purebred cattle. Nevertheless, when producers are losing money, they have fewer resources to invest in embryo transfer for purebreds than when they are making money. Thus, while the abrupt decrease in use of embryo transfer in 1976–1977 was due primarily to the sudden increase in numbers of "exotic" purebreds, it was exacerbated by low meat prices. Similarly, trends in the stock market affect embryo transfer because capital from investors serves as fuel.

There are also important tax considerations. Embryo transfer with breeding stock represents an excellent mechanism for converting ordinary income to long-term capital gains, which are taxed at much lower rates. Also, breeding stock qualifies for investment credit, which may be deducted directly from tax liabilities. Many embryo transfer costs are legitimately deducted as direct expense and can be balanced off against other profit-making enterprises. This works especially well if one is in the cattle or horse business both for pleasure and for profit.

Eventually, embryo transfer costs will be low enough and success rates high enough that this technology can profitably contribute to increased milk and meat production. There are only a few such situations to date, for example, getting several sons per year for artificial insemination from outstanding cows. In this context, embryo transfer technology has a much greater role to play when profits are reasonably high than when they are not (see Chapter 13, this volume).

It is, of course, difficult to predict future economic conditions, especially for agriculture. In the absence of severe recession, depression, or major wars, it is likely that the great world surplus of dairy products will cause some problems by the mid-1980s. Similarly, by mid-decade, we should expect the unfavorable portion of the beef–cattle cycle in North America (Miller *et al.*, 1979), although this may not follow historical trends due to rapidly increasing energy prices. These and related factors will certainly modulate the extent of embryo transfer use, although factors such as lower costs and increased efficacy may be more important.

B. Regulatory Agencies

Several organizations in the United States are concerned with various aspects of regulation of the embryo transfer industry. These include customs officials of the Treasury Department, the United States Congress (through legislative decisions, including allocation of Federal research funds), the Food and Drug Administration, the United States Department of Agriculture, especially through its division of Animal and Plant Health Inspection Service, animal regulatory arms of state governments, and breed associations. Similar groups exist in most countries. Their interest is to regulate commerce, to assure continued sources of animal protein, to prevent the spread of livestock diseases, to maintain human health by controlling drugs and procedures used in food producing animals, and to maintain or increase genetic progress in purebred animals.

Many new technologies are first applied commercially to purebred animals because the expense is more easily justified with animals of high value. This means that breed societies can exercise considerable influence on the rate of adoption of a technology in the critical period from the time when it becomes feasible (but expensive) to when it is more practical and costs less. For embryo transfer with cattle, this critical period will last for just over a decade, with costs (in noninflated constant dollars) declining by about an order of magnitude during this period.

Breed societies promote the development of a certain breed and the interests of society members. After the formative years of a society, for practical purposes, animals of the breed are defined as offspring from animals already recognized by the society because the parents are officially registered. Hence criteria for registration are not usually related to production traits. On balance, the record of breed societies has been to preserve the status quo, but they frequently do promote innovation. Cattle breed associations permit the registration of calves produced by embryo transfer (Seidel *et al.*, 1980), but usually require blood typing. Their concern is to ensure parentage more than to compare performance records, normality, or sex ratios of calves produced by embryo transfer to those produced conventionally. A few equine breed associations allow the registration of embryo transfer offspring, but usually only one foal per mare per year, and in some instances, the mare must either have been barren for a specified number of years or be above a certain age. In the horse breeding industry, the regulations of breed associations are designed more to preserve the prevailing market structure than to improve the breed.

The power of breed societies is so great that one must either bow to the rules or submit to the slow, painful, humbling, and frequently unsuccessful experience of trying to change them. If, for example, one were caught using artificial insemination with many breeds of horses, the consequences would probably be annulment of the registration papers of all animals involved (thus making the animals essentially worthless) and expulsion of the breeder from the breed society so that he (and frequently certain relatives) could never register animals of that breed. Sometimes these practices are challenged in court, but rarely successfully, in part because of the costs. Some years ago, however, a cattle breed association was successfully prosecuted on the rules against artificial insemination on the basis of the Sherman anti-trust laws.

Most activities of breed associations are entirely legitimate and focus on maintaining purity of the breed. One such function in the beef industry has been to assume responsibility for breed improvement with the aid of the Beef Improvement Federation. Another function is to verify that the parents of a registered animal are the true parents, usually by blood typing. In general, there are no routine blood-typing requirements for natural breeding, limited requirements for artificial insemination, and stringent requirements for embryo transfer. The stringency is related to the potential for fraud, but not to that for error. The chance for error is greatest for natural mating (about 5% from surveys) and least for embryo transfer (near 0%).

Professional societies and trade associations that play a role in the development and application of animal breeding technology are listed in Table III. They provide a forum for introduction and peer review of new methodology, a mechanism for the automatic collection and compilation of data and a link between research, industry, and professional practice. The National Association of Animal Breeders compiles records by breed of the number of units of semen processed and sold. Similarly, the Beef Improvement Federation gathers production data for beef cattle. The records are collected with no commercial pressure, just as data on sire performance and production of dairy cattle are gathered by the Dairy Herd Improvement Association (DHIA). The information gathered in this way is unbiased and reliable; in contrast, information gathered by vested interests is generally presented in a manner designed to promote those interests. It has been estimated that the service summarizing DHIA data has returned value on the investment at a rate of 20 to 1, for both the industry and the nation.

Conclusive documentation of the effect of embryo transfer on characteristics of the pregnancies and offspring could be obtained from the vast records accumulated by commercial embryo transfer centers if funds were available. The data have not been recorded in a uniform way, so comparisons cannot be made without first translating the statistics into common terms. Because the industry is highly competitive, efforts would have to be made to eliminate such distortions

TABLE III

United States Organizations Concerned with Animal Breeding Technology

1. American Society of Animal Science
2. American Association of Bovine Practitioners
3. American College of Theriogenologists (Board certified veterinarians specializing in reproduction)
4. American Dairy Science Association
5. American Society for Theriogenology
6. American Veterinary Medical Association
7. Beef Improvement Federation
8. Certified Semen Services
9. Council of Animal Scientists and Technologists
10. International Embryo Transfer Society
11. National Association of Animal Breeders
12. National Dairy Herd Improvement Association
13. Purebred Dairy Cattle Association

as would arise from secretiveness or the desire to show one's service to best advantage.

The International Embryo Transfer Society was formed in 1974 to facilitate the exchange of information among professionals interested in commercial embryo transfer in animals (predominantly cattle). The membership has grown to nearly 500 (regular, associate, and student members combined from the United States, Canada, and 22 other nations) in 1981 and represents interests not only in commercial service, but also in research, trade of embryos in both domestic and international markets, regulation of standards, and human reproduction. In addition to the dissemination of information through annual conferences and the publication of abstracts and bibliographies, the role of the society now encompasses the formulation of industry-wide recommendations on import–export regulations, standard codes for identification of frozen embryos, registration requirements for offspring produced by embryo transfer, the identification of research priorities, as well as quality control of member transfer units or practitioners within the United States.

The European Economic Community is keenly interested in the industry and has sponsored international conferences that have generated some of the major reference documents on embryo transfer technology (Rowson, 1976; Sreenan, 1978a). The focus on embryo transfer of the VIIIth and IXth International Congresses on Animal Reproduction and Artificial Insemination (Krakow, 1976 and Madrid, 1980), the Beltsville Symposium in Agricultural Research (1978), and the World Conference on Instrumental Insemination, *In Vitro* Fertilization, and Embryo Transfer (Kiel, 1980) demonstrates sustained interest in both research and commercial applications on an international scale.

Governmental regulations such as health regulations for moving cattle are essential for many aspects of the industry. Frequently, however, an inordinate amount of effort is required to cater to the regulations, especially those concerning new drugs. It has been reasonably easy to get approval for performing experiments with small numbers of animals to establish safety and efficacy (by obtaining a permit for "Investigation of a New Animal Drug," INAD), but the FDA has been appropriately conservative in requiring that the animals not be used for food purposes for long periods.

The real problem, however, is obtaining approval for routine use. First, an entrepreneur is required with huge financial resources, usually a drug company. Next, data on efficacy and safety are developed and submitted to the FDA. One curious aspect of this process is that the FDA has no clearcut criteria for such data. The drug company submits data, which it guesses will satisfy the FDA, who then consider the data and usually request more studies. Then these data are obtained, and the process repeated. This is not completely unreasonable, as the FDA does not have the resources to both design and evaluate appropriate experiments for all the drugs that come along. For a variety of reasons, the delays have added up to several years for the first efficacious estrus synchronization drug to be approved (prostaglandin $F_2\alpha$ produced by the Upjohn Company, approved for use in cattle in November, 1979). In all likelihood another prostaglandin produced by ICI, Ltd., will be approved in 1–2 years. The ICI drug has been approved for use for several years in many other countries including Canada.

Approval of the most efficacious drugs for estrus synchronization, progestins, is not being sought at present because an estrogen injection is also required, and the emotional situation is such that industry and scientific personnel believe that the FDA cannot approve the scheme, even if presented with reasonable data. Therefore, no one is willing to undertake the risk. That the estrogen is a genuine risk to health is extremely improbable because (1) tissues of mammals (including males) have estrogens in them already since estrogens are essential for reproduction, (2) the injection is given nearly 2 weeks before breeding, which in nearly all cases will be months to years before animals are used for meat (milk cows are a less clear case), and (3) the kind of estrogen given is essentially inactive orally, so eating it accidentally would have little effect. Thus, the issue is emotional and political, not scientific or technological. Although drugs for estrus synchronization have been used as an example, the same problems apply to drugs for superovulation. It is to be hoped that it will be possible to avoid additional excessive regulations in other areas such as culture media.

C. Improved Technologies

While the technology of embryo recovery and transfer may have limited potential for further improvement, many related technologies will improve markedly in the coming two decades, resulting in a reduction of costs and much wider application. The following predictions seem appropriate for the cattle embryo transfer industry:

1. By the end of 1982, 80–90% of embryo transfer will be done nonsurgically.

2. By the end of 1982, 50% of the embryos recovered will be frozen, increasing to 80–90% by the end of 1983. The cost per embryo for freezing will be $50–$100.

3. By the end of 1981, 70% of transfers will be on-farm, increasing to 80–90% by the end of 1982.

4. Sex determination of embryos will develop slowly but surely and will be applied to 10–20% or more of embryos transferred by 1983.

5. By the end of 1983, there will be more than 1000 technicians trained to thaw, evaluate, and transfer embryos. Fees will be $25 (based on the value of the dollar in 1980) per transfer if provided by the regular artificial insemination technician and $100 plus travelling expenses plus $25 per transfer outside of

areas serviced by such technicians. Costs of embryos themselves will be additional.

6. By 1982–1983, demand for superovulation of donors for freezing embryos will be high; entrepreneurs will be ready so profits will not be excessive. Trade in frozen embryos will be active.

7. In 1984 or 1985, the rate of increase of embryo transfer services will level off (Table IV) as it becomes clear that embryo transfer is not a very potent genetic tool compared to artificial insemination. Demand will again increase in the last half of the decade as people learn where these techniques really can be useful.

8. There is great potential for exciting developments in the 1990s, such as cloning, but it is impossible to predict accurately what will evolve and when.

Coupled to improvements in herd management, superovulation, estrus synchronization, cryopreservation of embryos, nonsurgical transfer, and marketing will be improvements in new technologies that have not yet been applied to the livestock industry. These might include induced twinning, sex selection and determination, *in vitro* fertilization, production of identical twins, crossing of two males, manufacture of homozygous individuals within a single generation, and cloning (see Chapter 10, this volume). Such technologies, once developed for practical application, promise more rapid genetic progress, wider opportunities for selection and control over breeding, greater alleviation of infertility, and increased production of animal protein. There are no known biological barriers to the eventual development of any of these technologies; the speed with which they become available will depend in large measure on the economic incentives. Of course, technologies such as these can never come into being without new information from basic research. Adoption of such new technologies will increase the scope of the embryo transfer industry greatly, but its application will remain too specialized and too expensive to equal the volume of the artificial insemination industry in the next two decades.

TABLE IV

Estimated and Predicted Pregnancies from Bovine Embryo Transfer (North America)

Year	Number
1978	10,000
1980	25,000
1982	50,000
1984	100,000
1986	150,000
1988	200,000
1990	250,000

D. Popular Opinion

Favorable popular opinion always influences the rate at which a new technology is adopted. In the case of embryo transfer, it has actually provided the impetus to the development of the industry. Although the volume of commercial embryo transfer has been documented only for cattle, the following statistics show how rapidly the industry has grown. Prior to 1972, there was essentially no commercial embryo transfer activity; in 1979, an estimated 13,000 embryos were transferred in the United States and nearly 4000 in Canada (Seidel, 1981a). In 1974, there were two calves produced by embryo transfer registered with the American Holstein–Friesian Association compared to 1790 in 1980 (Nelson, 1981). This rapid adoption of a new and relatively unproven technology by traditionally conservative livestock breeders contrasts sharply with the development of the artificial insemination industry (see Chapter 2, this volume). This is due in part to the fact that investors who do not make their living by breeding livestock have been attracted by the prolific and favorable popular and trade press coverage that embryo transfer has received (S. Seidel, 1980). Funds from these investors, who make their profit primarily in tax benefits and promotion, have supported much of the research to develop the industry. The opportunities for this symbiotic relationship will grow as the frozen-embryo market expands.

E. Identification of the Greatest Needs

The National Association of Animal Breeders, the International Embryo Transfer Society, and the United States Department of Agriculture (Waters, 1981) have all expressed the concern that the embryo transfer industry, particularly the bovine portion of it, has not yet adopted a standardized code for the permanent identification of frozen embryos. The expanding international trade in embryos and rapidly improving freezing techniques make this perhaps the most critical need at this point, although a relatively easy one to meet. The information with which each frozen embryo must be permanently identified includes (1) registration numbers or other permanent identification of the dam and sire, (2) age of the embryo measured in number of days since the donor was first observed in standing estrus, (3) date and method of freezing, since the method of thawing is determined by the method of freezing, and (4) the name of the company or individual who processed the embryo, so that other significant information can be obtained. The other records that must be kept pertain to treatment of the donor (e.g., superovulatory drugs), morphology and other data characterizing the embryo, health status of the donor and sire, and the composition of collection and storage media. The latter is necessary since most media contain heat-inactivated serum or serum albumin, which may be carriers of certain microorganisms that are localized to a geographical or political area (Waters, 1981).

The embryo transfer industry has also perceived the need to police itself to assure that high standards of ethics and practices be fostered. To this end, the United States members of the International Embryo Transfer Society are formulating a means to certify member embryo transfer units and practitioners. This will provide some definition of acceptable practices and will give clients a more objective basis on which to select firms to provide embryo transfer and related services. Canada already has a rudimentary system for certifying embryo transfer centers. To an extent, the Holstein–Friesian Association of America has already met this need in North America for customers who purchase frozen embryos by organizing a commercial subsidiary to (1) identify cows and sires on the basis of performance records that are acceptable for embryo transfer, (2) to negotiate the terms of purchase including establishing a price scale, and (3) to identify practitioners to carry out the superovulation, recovery, processing, and eventual transfer of the embryos (Nelson, 1981). The Association has also developed a number of technical assistance programs for countries outside of the United States to provide training and expertise necessary to develop markets for commerce in livestock embryos.

One thorny issue that has not been resolved in North America is whether commercial embryo transfer procedures are unlawful when done in the absence of the direct supervision of a veterinarian. Laws governing licensing and conduct of the practice of veterinary medicine vary somewhat from state to state, and thus far, the authorities and veterinary associations have taken little notice of possible infringements. Recently, Australia made it illegal for these procedures to be carried out by nonveterinarians. Obviously, an appropriate Master's or Ph.D. degree program can provide better training for the technical conduct of some of these very specialized procedures than a degree in veterinary medicine without postgraduate specialization in theriogenology. Many procedures can now be done entirely nonsurgically and, therefore, are akin to artificial insemination. For both of these technologies, much variability in results is attributable to abilities of individual operators (Seidel, 1981b). Manual facility to perform these tasks improves with experience, and frequent practice seems important. As with artificial insemination, veterinarians and others with professional training will predictably and appropriately play a less active direct role as noninvasive procedures become routine.

An extremely complex regulatory need concerns health regulations for moving embryos from one area to another. Unfortunately, the interaction between embryos and pathogens has not been determined for more than a few diseases (Waters, 1981), and intelligent regulations are difficult to formulate without knowing what diseases can be transmitted by embryos. Further information about mechanisms of transmission would be extremely desirable (Waters, 1981), and in some cases, would lead to prevention of transmission.

Regulations to date have been formulated by using those for movement of live animals as a starting point, and sometimes adding additional restrictions. It seems incongruous to have more restrictive regulations for embryos than for mature animals, especially for movement between contiguous areas such as Canada and the United States, since the handling of embryos intrinsically involves safety factors, such as dilution of microorganisms with large volumes of sterile media, antimicrobial agents in the media, the impossibility of transmitting many diseases by the intrauterine route, the absence of receptors for some viruses on embryos, and the protective barrier of the zona pellucida. Although one could perhaps imagine a case in which disease could be transmitted by embryos but not by the mature animal with an embryo (i.e., pregnant), this seems very unlikely. Therefore, if the donor meets the health regulations for moving animals, it seems logical that her embryos should also qualify.

The International Embryo Transfer Society has identified research priorities for the embryo transfer industry (Table V) as a result of surveys (Godke, 1981) conducted among the membership. The rankings were developed from opinions of about 165 members returning surveys, and therefore, probably represent perceived needs of the industry quite accurately.

Nearly all applications of embryo transfer technology depend on the use of gonadotropins and prostaglandins or other drugs to synchronize estrus. There is a need for competitive sources of these drugs and for better technology and more strictly applied measures of quality control to ensure sufficient supplies of standardized products. It is likely that some of the variability in response to gonadotropins could be eliminated in this manner.

Very few veterinary curricula have included instruction in embryo transfer methods; thus, adequate training has been difficult to obtain. In the United States, some universities offer graduate training programs in embryo transfer, and there are several short courses offered to veterinarians or breeders to give orientation but by no means skills enough to carry out successful embryo transfer. For the most part, the only training has been on-job experience with commercial firms. There is an immediate need for established training programs to turn out competent, well-qualified technicians, and the need will grow along with the trend toward nonsurgical transfer of frozen embryos on the farm.

TABLE V

Ranking of Research Priorities from Surveys of the Members of the International Embryo Transfer Society

1. Cryopreservation of embryos
2. Nonsurgical embryo transfer (horses and cattle)
3. Superovulation and follicular development
4. Evaluation of viability of embryos
5. Fertilization
6. Sex ratio control; sex determination of embryos
7. *In vitro* culture and manipulation of embryos
8. Uterine environment and embryo metabolism

One final need, one which may not be so easy to meet, is for the social wisdom to use embryo transfer and resulting new technologies for good. While technology is at hand to counter some of the ill effects of growing population and declining agricultural and energy resources on the availability of animal protein, the political means to adopt it may not be.

IV. SUMMARY

The embryo transfer industry has become firmly established in more than a dozen countries. The vast majority of commercial activity concerns cattle, although work with horses is increasing rapidly. Commercial embryo transfer with swine, sheep, goats, and rabbits is more limited, but important in some instances. Information generated through these procedures in domestic animals will undoubtedly contribute to alleviating reproductive problems in man.

Although there is clear documentation (e.g., Seidel, 1981a) of a large and thriving industry, one might legitimately question the long-term value of much of the work. Suspicions are aroused because meat and milk production are frequently secondary or even tertiary considerations when selecting cattle donors. That commercial work is essentially limited to purebred animals, which are frequently owned by millionaires who treat the whole enterprise as a hobby, is somewhat distressing. That so much emphasis is on tax considerations is worrisome. That many people have lost huge sums of money is of concern, even though they frequently brought it on themselves. Yet the situation has many positive aspects. More than 1000 families make their living, directly or indirectly, by providing embryo transfer services. Currently, most people who contract these services are full-time farmers and ranchers, and it must be admitted that the entrepreneurs, businessmen, and hobbyists have by and large been a positive force, especially in establishing the industry. The research value of these techniques has tremendous potential (Seidel, 1979), especially in improving reproductive efficiency outside the context of embryo transfer. Moreover, one can confidently look forward to inexpensive and efficacious techniques that can be applied on a cost-effective basis to increased productivity of animals. Possibly, pregnancy rates with nonsurgical embryo transfer will someday be higher than with artificial insemination.

Furthermore, the clearly positive accomplishments already achieved should not be ignored. These include (1) a great broadening of the genetic base of certain breeds of cattle in countries where only a few individuals existed, (2) the bulls proven to be free of certain genetic defects, (3) the joy of obtaining a large litter of calves from a valuable cow with a senescent uterus, (4) the efficacious importation of genetic material to increase milk production in some countries, and (5) the introduction of new genes into a swine herd without compromising its pathogen-free status, and (6) the birth of Louise Brown.

There is much more to come. Fruition of embryo transfer technology is a triumph of the human spirit as much as it is an economically sound enterprise. To most people, the concept of tricking nature in this way is immediately appealing in some intrinsic way analogous to enthusiasm for music or sports. Those who have worked in this area know the problems of fitting the many pieces together which culminate in success or failure. People who do embryo transfer must rely absolutely on one another, because one person cannot do everything—the estrus detection, feeding, sorting, hauling, vaccinating, superovulation, insemination, embryo recovery, embryo evaluation, embryo storage, embryo transfer, pregnancy diagnosis, record keeping, buying, selling, and billing. Clearly, successful superovulation and embryo transfer involve much plain hard work, be it with mice or cows.

Quite a few individuals have now seen thousands of mammalian embryos. Somehow, even with the most experienced embryologists, there is a special feeling when 15 beautiful bovine embryos from one donor are seen under a microscope when nature originally intended only one—and that not to be exposed to the eye of man. The disappointment of 28 unfertilized bovine ova from a valuable donor is also frequently registered.

The long-term value of embryo transfers will depend in great part on fruition of the technologies discussed in other chapters of this book, and their success will depend on improved embryo transfer. Sexing, cloning, freezing, genetic engineering, consistently high rates of success with simple, nonsurgical embryo transfer in various species, and many other achievements are worthy of continued efforts and monetary support. On a world-wide basis, it has got to be a better investment than weapons.

ACKNOWLEDGMENTS

Numerous friends and colleagues have contributed ideas and information appearing in this chapter. The authors especially acknowledge the help of R. P. Elsden and E. L. Squires. Our research has been supported in part by the Colorado State University Experiment Station through Regional Project W-112, Reproductive Performance in Cattle and Sheep.

REFERENCES

Allen, W. R. (1977). *In* "Embryo Transfer in Farm Animals" (K. J. Betteridge, ed.), Monograph 16, pp. 47–49. Canada Department of Agriculture, Ottawa.
Anderson, G. B. (1978). *Theriogenology* **9,** 3–16.
Anderson, L. L. (1980). *In* "Reproduction in Farm Animals" (E. S. E. Hafez, ed.), pp. 358–386. Lea & Febiger, Philadelphia, Pennsylvania.
Anonymous (1980a). *Hoard's Dairyman* **125,** 1555.
Anonymous (1980b). *J. Am. Vet. Med. Assoc.* **178,** 106.
Baker, R. D., Snider, G. W., Leipold, H. W., and Johnson, J. L. (1980). *Theriogenology* **13,** 87.

Beauprez, R. (1980). *Proc. 6th IETS Owners Managers Workshop, Denver 1980*, pp. 21-25.

Bedirian, K. N., Mills, M. S., Bligh, P. J., Geroldi, R., and Kilmer, B. A. (1979). *Theriogenology* **11**, 3-4.

Betteridge, K. J. (1977). "Embryo Transfer in Farm Animals," Monograph 16. Canada Department of Agriculture, Ottawa.

Betteridge, K. J. (1981). *J. Reprod. Fertil.* **62**, 1-13.

Betteridge, K. J., and Moore, N. W. (1977). *In* "Embryo Transfer in Farm Animals" (K. J. Betteridge, ed.), Monograph 16, pp. 37-38. Canada Department of Agriculture, Ottawa.

Betteridge, K. J., Eaglesome, M. D., Randall, G. C. B., and Mitchell, D. (1980). *J. Reprod. Fertil.* **59**, 205-216.

Bowen, R A., Elsden, R. P., and Seidel, G. E., Jr. (1978). *J. Am. Vet. Med. Assoc.* **172**, 1303-1306.

Brackett, B. G., Oh, Y. K., Evans, J. F., and Donawick, W. J. (1980). *Biol. Reprod.* **23**, 189-205.

Brand, A., and Drost, M. (1977). *In* "Embryo Transfer in Farm Animals" (K. J. Betteridge, ed.), Monograph 16, pp. 31-34. Canada Department of Agriculture, Ottawa.

Brem, G. (1979). Ph.D. Dissertation, Ludwig-Maximilians-Universität, Munich.

Bulman, D. C., and Lamming, G. E. (1979). *Br. Vet. J.* **135**, 559-567.

Caras, R. (1980). *GEO* **2**, 80-94.

Carmichael, R. A. (1980). *Theriogenology* **13**, 3-6.

Christie, W. B., Newcomb, R., and Rowson, L. E. A. (1979). *Vet. Rec.* **104**, 281-283.

Christie, W. B., Newcomb, R., and Rowson, L. E. A. (1980). *Vet. Rec.* **106**, 190-193.

Curtis, J. L., Elsden, R. P., and Seidel, G. E., Jr. (1981). *Theriogenology* **15**, 124 (Abstr.).

Day, B. N. (1979). *Theriogenology* **11**, 27-31.

Douglas, R. H. (1979). *Theriogenology* **11**, 33-46.

Durrant, B., and Benirschke, K. (1981). *Theriogenology* **15**, 77-83.

Eller, J. (1981). *Proc. 7th IETS Owners Managers Workshop, Denver 1981*, pp. 4-15.

Elsden, R. P., Hasler, J. F., and Seidel, G. E., Jr. (1976). *Theriogenology* **6**, 523-532.

Elsden, R. P., Nelson, L. D., and Seidel, G. E., Jr. (1978). *Theriogenology* **9**, 17-26.

Elsden, R. P., Nelson, L. D., and Seidel, G. E., Jr. (1979). *Theriogenology* **11**, 17-26.

Elsden, R. P., Nelson, L. D., Case, L. G., and Seidel, G. E., Jr. (1980). *Theriogenology* **13**, 95.

Foote, R. H., and Onuma, H. (1970). *J. Dairy Sci.* **53**, 1681-1692.

Godke, R. A. (1981). *April Newsletter, Int. Embryo Transfer Soc.*, **5**.

Gordon, I. (1975). *Ir. Vet. J.* **29**, 21-30, 39-62.

Gordon, I. (1977). *Proc. Symp. Management Reprod. Sheep Goats, Madison 1977*, pp. 15-30.

Greve, T., and Lehn-Jensen, H. (1979). *Acta Vet. Scand.* **20**, 135-144.

Guthrie, H. D., and Polge, C. (1976). *J. Reprod. Fertil.* **48**, 427-430.

Halley, S., Rhodes, R. C., III, McKellar, L. D., and Randel, R. D. (1979). *Theriogenology* **12**, 97-108.

Hansel, W., and Beal, W. E. (1979). *In* "Animal Reproduction, BARC Symposium 3" (H. Hawk, ed.), pp. 91-110. Allanheld, Osmun, Montclair, New Jersey.

Hare, W. C. D., and Betteridge, K. J. (1978). *Theriogenology* **9**, 27-43.

Harlow, G. M., and Quinn, P. (1979). *Aust. J. Biol. Sci.* **32**, 363-369.

Heape, W. (1891). *Proc. R. Soc. London* **48**, 457-458.

Hirao, Y., and Yanagimachi, R. (1978). *J. Exp. Zool.* **206**, 365-369.

Holtan, D. W., Douglas, R. H., and Ginther, O. J. (1977). *J. Anim. Sci.* **44**, 431-437.

Imel, K. J., Squires, E. L., and Elsden, R. P. (1980). *Theriogenology* **13**, 97 (Abstr.)

Imel, K. J., Squires, E. L., Elsden, R. P., and Shideler, R. K. (1981). *J. Am. Vet. Med. Assoc.* (in press).

Jillella, D., and Baker, A. A. (1978). *Vet. Rec.* **103**, 574-576.

Johnson, J. L., Leipold, H. W., Snider, G. W., and Baker, R. D. (1980). *J. Am. Vet. Med. Assoc.* **176**, 549-554.

Karihaloo, A. K. (1981). *Proc. 7th IETS Owners Managers Workshop, Denver 1981*, pp. 27-30.

Kummerfeld, H. L., Oltenacu, E. A. B., and Foote, R. H. (1978). *J. Dairy Sci.* **51**, 1773-1777.

Kunkel, R. N., and Stricklin, W. R. (1978). *Theriogenology* **9**, 96 (Abstr.).

Leidl, W., Himmer, B., and Fung, H. (1980). *Prakt. Tierarzt* **61**, 6-9.

Lubbadeh, W. F., Graves, C. N., and Spahr, S. L. (1980). *J. Anim. Sci.* **50**, 124-127.

Markette, K. L., Seidel, G. E., Jr., and Elsden, R. P. (1980). *Theriogenology* **13**, 103 (Abstr.).

Maurer, R. R., and Foote, R. H. (1971). *J. Reprod. Fertil.* **25**, 329-341.

Miller, W. C., Rossiter, D. L., Ward, G. M. and Yorks, T. P. (1979). *J. Anim. Sci.* **49**, 629-636.

Moore, N. W. (1977). *In* "Embryo Transfer in Farm Animals" (K. J. Betteridge ed.), Monograph 16, p. 40. Canada Department of Agriculture, Ottawa.

Nelson, L. D., Bowen, R. A., and Seidel, G. E., Jr. (1975). *J. Anim. Sci.* **41**, 371-372 (Abstr.).

Nelson, L. D., Elsden, R. P., Homan, N. R., and Seidel, G. E., Jr. (1980). *Theriogenology* **13**, 106. (Abstr.).

Nelson, R. E. (1981). *Proc. 7th IETS Owners Managers Workshop, Denver 1981*, pp. 31-42.

Newcomb, R. (1979). *Vet. Rec.* **105**, 432-434.

Newcomb, R., Christie, W. B., and Rowson, L. E. A. (1978). *J. Reprod. Fertil.* **52**, 395-397.

Polge, C. (1977). *In* "Embryo Transfer in Farm Animals" (K. J. Betteridge, ed.), pp. 43-45. Canada Department of Agriculture, Ottawa.

Renard, J.-P., Heyman, Y., and Ozil, J.-P. (1980a). *Vet. Rec.* **107**, 152-153.

Renard, J.-P., Philippon, A., and Menezo, Y. (1980b). *J. Reprod. Fertil.* **58**, 161-164.

Rowe, R. F., Del Campo, M. R., Critser, J. K., and Ginther, O. J. (1980a). *Am. J. Vet. Res.* **41**, 106-108.

Rowe, R. F., Del Campo, M. R., Critser, J. K., and Ginther, O. J. (1980b). *Am. J. Vet. Res.* **41**, 1024-1028.

Rowson, L. E. A. (1976). "Egg Transfer in Cattle." European Economic Community, Luxembourg.

Rowson, L. E. A., Moor, R. M., and Lawson, R. A. S. (1969). *J. Reprod. Fertil.* **18**, 517-523.

Rowson, L. E. A., Lawson, R. A. S., and Moor, R. M. (1971). *J. Reprod. Fertil.* **25**, 261-268.

Scaramuzzi, R. J. (1979). *J. Steroid Biochem.* **11**, 957-961.

Scaramuzzi, R. J., Davidson, W. G., and VanLook, P. F. A. (1977). *Nature (London)* **269**, 817-818.

Scaramuzzi, R. J., Baird, D. T., Martensz, N. D., Turnbull, K. E., and Van Look, P. F. A. (1981). *J. Reprod. Fertil.* **61**, 1-9.

Schilling, E., Niemann, H., Chang, S. P., and Doepke, H.-H. (1979a). *Zuchthygiene* **14**, 170-172.

Schilling, E., Smidt, D., Sacher, B., Petac, D., and El Kaschab, S. (1979b). *Ann. Biol. Anim. Biochim. Biophys.* **19**, 1625-1629.

Schneider, H. J., Jr., Castleberry, R. S., and Griffin, J. L. (1980). *Theriogenology* **13**, 73-85.

Schneider, U., and Hahn, J. (1979). *Theriogenology* **11**, 63-80.

Schultz, R. H. (1980). *Theriogenology* **13**, 7-11.

Seidel, G. E., Jr. (1977). *In* "Embryo Transfer in Farm Animals" (K. J. Betteridge, ed.), Monograph 16, pp. 20-24. Canada Department of Agriculture, Ottawa.

Seidel, G. E., Jr. (1979). *In* "Animal Reproduction, BARC Symposium 3" (H. Hawk, ed.), pp. 195-212. Allanheld, Osmun, Montclair, New Jersey.

Seidel, G. E., Jr. (1980a). *Guernsey Breeders' J.* **146**, 16-18.

Seidel, G. E., Jr. (1980b). *Proc. Bovine Embryo Transfer Short Course, Colorado State Univ.* 6 pp.

Seidel, G. E., Jr. (1981a). *Science (Washington, D.C.)* **211**, 351-358.

Seidel, G. E., Jr. (1981b). *In* "Fertilization and Embryonic Development *In Vitro*" (L. Mastroianni, J. D. Biggers, and W. Sadler eds.), Plenum, New York (in press).

Seidel, G. E., Jr., and Seidel, S. M. (1978). *Adv. Anim. Breeder* **26**, 6-10.

Seidel, G. E., Jr., Seidel, S. M., and Bowen, R. A. (1978, revised 1980). "Bovine Embryo Transfer Procedures." Colorado State Univ. Exp. Stn. General Series 975.

Seidel, S. M. (1980) *April Newsletter. Int. Embryo Transfer Soc.* pp. 13-16.

Seidel, S. M., and Seidel, G. E., Jr. (1979, 1980). *Theriogenology* **12, 13, 14, 14,** 339–410, 237–248, 227–238, 477–492.

Shea, B. F. (1981). *Theriogenology* **15,** 31–42.

Sreenan, J. M. (1978a). "Control of Reproduction in the Cow." Martinus Nijhoff, The Hague.

Sreenan, J. M. (1978b). *Theriogenology* **9,** 69–83.

Sreenan, J. M., and McDonaugh, T. (1979). *J. Reprod. Fertil.* **56,** 281–284.

Sugie, T., Seidel, G. E., Jr., and Hafez, E. S. E. (1980). *In* "Reproduction in Farm Animals" (E. S. E. Hafez, ed.), pp. 569–594. Lea & Febiger, Philadelphia, Pennsylvania.

Tervit, H. R., Cooper, M. W., Goold, P. G., and Haszard, G. M. (1980). *Theriogenology* **13,** 63–71.

Tischner, M., and Bielanski, A. (1980). *J. Reprod. Fertil.* **58,** 357–361.

Trounson, A. O., Willadsen, S. M., and Moor, R. M. (1976). *J. Agric. Sci.* **86,** 609–611.

Van Look, P. F. A., Clarke, I. J., Davidson, W. G., and Scaramuzzi, R. J. (1978). *J. Reprod. Fertil.* **53,** 129–130.

Waters, H. A. (1981). *Theriogenology* **15,** 57–66.

Wright, J. M. (1981). *Theriogenology* **15,** 43–56.

Wright, R. W., Jr., Anderson, G. B., Cupps, P. T., and Drost, M. (1976a). *J. Anim. Sci.* **43,** 170–174.

Wright, R. W., Jr., Anderson, G. B., Cupps, P. T., Drost, M., and Bradford, G. E. (1976b). *J. Anim. Sci.* **42,** 912–917.

4

New Aspects of Poultry Breeding

THOMAS J. SEXTON

I. INTRODUCTION

Considerable advancement has been made in poultry breeding technology in the last 30 years, but significant improvements in the efficiency of production may still be made through new technologies. Nearly all of the breeding behind the primary poultry types (egg-production chickens, meat-production chickens, and turkeys) is carried out by commercial breeding organizations seeking profits on the sale of breeding stock in the form of either parent breeders, baby chicks or poults, or ready-to-lay started pullets for egg production. Most of the commercial layers, broilers, and turkeys are produced by some system of crossbreeding. There are two general approaches used by primary breeders to improve the stock they sell: (1) to develop and evaluate new strains to replace old strains going into the crossbreeding system, or (2) to select within each of the strains of the crossbreeding system so as to improve the efficiency of performance of the

NEW TECHNOLOGIES IN ANIMAL BREEDING

progeny. Regardless of the breeding system selected, the objectives of any system will be: (1) to increase egg production at sizes and quality characteristics desired by the consumer and (2) to continue the emphasis, with broilers and turkeys, upon growth rate, feed efficiency, yield, reduction in body fat, and freedom from defects leading to condemnations.

Most of the economically important highly heritable traits such as growth rate, body conformation, sexual maturity, egg weight, and shell quality are perpetuated by mass selection because there is little advantage in hybrid vigor. Low heritable traits such as egg production, livability, fertility, and disease resistance are perpetuated by crossbreeding and identified through progeny testing and family testing.

The primary purpose of this chapter is to outline the present technologies, both genetic and reproductive, that are being used in commercial poultry breeding. A secondary objective is to discuss the application of potentially new technologies and those constraints that could alter their acceptance by the industry.

II. QUANTITATIVE GENETIC TECHNOLOGIES

The quantitative genetic technologies established in the late 1940s are still being used by the commercial geneticists. In a recent review by Sheldon (1980), it was pointed out that during the last 40 years quantitative genetic technologies have steadily advanced. More efficient methods have developed for estimating genetic parameters (heritability and genetic correlation) and their errors of estimate for the use of control populations, for analyzing genotype–environment interaction, for developing specialized sire and dam lines, for deciding the optimum structure of a breeding population in order to maximize the rate of progress, for constructing and using selection indices to study the effects of inbreeding, chance, and linkage, for estimating selection limits, for studying and utilizing nonadditive genetic variation, for studying the relationship of genetic distance between populations, and for incorporating cost–benefit factors into decision making based on animal breeding theory.

III. USE OF QUALITATIVE GENES

Historically, qualitative genes have provided the basis for the establishment of different varieties of poultry (Shaklee, 1974). These include, among others, plumage color, skeletal variation, and sex-linked barring. Most of the qualitative genes, which have been identified, have had deleterious effects when considered in terms of economic production, e.g., perosis (slipped tendons). However, several qualitative genes have proved to be economically beneficial. The autosomal dominant white feather gene (I) was used to develop the modern broiler strain. Another example is the dominant gene for white skin. Also the sex-linked barring gene (b) with its characteristic pattern effect on feather pigment has been widely used in auto-sexing breeds to determine sex at hatching time. The sex-linked rapid–slow feathering gene (k) has received more recent use for the same purpose (Marble and Jeffrey, 1955). Feather sexing is currently being used on a small scale by broiler (meat) producers. It is estimated that about 10% of the producers are feather sexing, whereas 10–15 years ago over 60% of the producers used feather sexing. The up-and-down interest in feather sexing by commercial broiler producers depends on two factors—market availability and production costs. The remaining broiler producers continue to raise broilers as straight-run chicks (mixed sexes). Sexing offers a distinct advantage in improving production because the males are faster growers than females and efficient management regimes can be designed accordingly. Feather sexing also has an advantage over the conventional method—examination of the rudimentary sex organs by highly trained personnel—because it is faster (1000 chicks/hr) and less expensive.

Among the known dwarfing genes, the recessive sex-linked dw gene (Hutt, 1949) is, with its alleles, the only gene that reduces body size without any marked detrimental effect upon reproductive ability and livability. From a practical point of view, dwarf poultry can be reared either as commercial egg layers or as meat-type breeder hens (mated to normal roosters). The first commercial crosses of dwarf breeders were sold in France in 1968 (for review, see Guillaume, 1976). It has been estimated that the dwarf layer and the dwarf broiler breeder hen will reduce production costs, when compared to their normal counterparts, by 20 and 2%, respectively.

IV. REPRODUCTIVE TECHNOLOGIES

There have been several reproductive technologies such as artificial insemination and semen preservation that have aided or accelerated the advances made through quantitative genetic technology. The practice of artificial insemination (AI) is widely used commercially by both primary and secondary breeders of turkeys because modern strains are heavy and broad-breasted, and this body conformation interferes with the ability to mate, which results in fertility of less than 50%. The advantages (Lake, 1978) of AI to the primary breeders are: (1) to improve the efficiency and accuracy of progeny testing or controlled breeding tests, (2) to increase selection pressure on the male line and hence, to accelerate the rate of genetic improvement, (3) to reduce the number of males used for breeding, and (4) to inseminate a greater number of females with semen from one male in a selective breeding program. The secondary breeders (purchasers of breeding stock for the purpose of reproducing birds for meat and egg production) use AI to (1) overcome the physical incompatibility between the sexes, (2) control the spread of reproductive pathogens, (3) overcome periods of low semen production caused by disease, faulty management practices, or severe climatic conditions, and (4) allow a more efficient use of males from farm to farm. Although some 3.5 million turkey hens are artificially bred annually (U.S. Department of Agriculture, 1978) less than one-half are inseminated with diluted semen. Why certain commercial producers have not adopted the use of semen extenders in their breeding programs is difficult to explain. Certainly those producers using extenders are reducing AI costs and at the same time maintaining fertility in flocks equal to flocks being inseminated with undiluted semen.

Despite the widespread use of AI by turkey breeding companies, application to chicken breeding has been limited to pedigree mating at the primary breeder level. In a recent survey, it was estimated that only 10% pedigree testing conducted by primary breeders of chickens was with AI. A number of factors have contributed to the lack of AI usage by primary or secondary breeders (Lake, 1978): (1) the high labor costs for artificially breeding approximately 387 million hens annually (U.S. Department of Agriculture, 1978), (2) the need for highly skilled labor to achieve good results, and (3) the lack of a serious study of the economic potentials.

Since 1948, considerable research has been devoted to the problem of semen storage (for reviews, see Lorenz, 1969; Sexton, 1979). The most significant breakthrough in poultry semen preservation occurred in 1967, when Van Wambeke of Belgium reported that diluted chicken semen could be held at 2°C for up to 24 hr without a significant loss in fertilizing capacity. Surprisingly few investigations have been conducted on the storage of turkey semen, although stored turkey semen has more potential commercial usefulness than preserved chicken semen. The optimal temperature for storing turkey semen is 15°C; even at this temperature, however, preservation without serious loss of fertilizing capacity has usually been limited to 6–8 hr. Despite these advances in preservation of semen at temperatures above 0°C, very little commercial application of these techniques has occurred. Until there is an increase in the use of AI in breeding chickens, methods for preserving chicken semen at temperatures above 0°C will be only a matter of curiosity to the research scientist. Unlike chicken semen, the level of fertility obtained with stored turkey semen is below that of unstored semen (80 versus 95%), and this differential is unsuitable by commercial standards.

The potential advantages of long-term preservation of poultry semen by freezing have long been recognized by researchers and breeders. Several semen freezing techniques have been reported that differ in methodology (e.g., diluents, cryoprotectants, and/or freezing methods). Only one, the Beltsville Method (Sexton, 1979), has been field tested and incorporated into the programs of several primary breeders. Although the level of fertility with frozen–thawed chicken semen (65%) is less than that with unfrozen semen (95%), the use of frozen semen in primary breeding programs offers the following advantages: (a) a way to increase genetic advancement per generation, (b) a means to gain a more precise evaluation of genetic improvement, (c) a way to allow more efficient use of sires, and (d) a way to distribute genetic material worldwide.

Recently, an encouraging degree of fertility (50%) has been obtained with frozen turkey semen (Sexton, 1979). However, further improvements are needed before field testing in turkeys can be considered.

V. FUTURE TECHNOLOGIES

To use the vernacular, there is both "good news and bad news" for future developments in poultry breeding technology. The good news is that the amount of genetic variation available in the foundation breeding stock has not diminished, and it is not expected to diminish in the near future. Of course, a genetic ceiling for certain traits such as growth rate will eventually be reached, but certainly not in the 1980s. Future poultry breeding at the primary level with layers, which will be less dramatic than in the past, will probably fall into three categories. First, efforts will continue in the maintenance and improvement of highly developed genetic lines and commercial crosses. Second, more emphasis will be placed on increasing total eggs marketed per hen with associated gains in feed efficiency and parallel gains in egg quality. Third, another future category will be in new product development, including synthesis of new genetic lines and improvement of reserve lines and crosses to meet specific present or future needs.

Growth rate (average weight/bird) in broilers will continue to increase at 4%/year. During the same period feed conversion will improve at about 1.5%/year. Extrapolation of these percentages suggests that birds will be reaching 4.4 lb by 5 weeks in the 1990s. The available selection pressure will be increasingly allocated to eviscerated yield, feed conversion, grade, and carcass composition. Breeding for physiological tolerance will become increasingly important, not only because of the increased use of intensive production systems but also to meet the increased physiological barriers resulting from faster growing and heavier broilers.

Turkey breeding in the last couple of decades has been characterized by the development of lines with lower cost per pound of meat and the development of male lines to add between 10 and 15% to the weight of the commercial products by transmitting better growth rate and meat yield. In the future, major advances will be made in the female lines by improving both egg production and meat yield.

Reproductive efficiency will continue to increase. Certain commercial breeders will place some emphasis on developing crosses adapted to cage housing. Such crosses will promote the use of management systems that are designed with labor and other efficiencies in mind. Disease control will be made easier at the commercial level by production of specific pathogen-free breeding stock (i.e., stock free of *Mycoplasma gallisepticum, M. synoviae,* and *M. meleagridis*). This advancement has recently been made by several breeders and is being worked on by others.

The commercial producers (integrators) throughout the world will insist that the primary breeder should only concentrate on basic genetic breeding. The integrators will plan on growing their own parent stock for hatchery supply flocks and having their own male and female lines. The rate of genetic progress made by the commercial breeders in broiler traits may be fully competitive with the major primary breeders. They will establish their own product image by emphasizing specific quality criteria in the final products. In particular, integrators will emphasize selection programs to reduce the days to market, improve feed conversion ratio, decrease mortality, and reduce the incidence of leg problems. A number of single genes and lines with specific traits are potentially useful for the commercial breeder: roaster lines, dwarf female lines, white skin male lines, male and female lines to produce color-sexable broilers, lines to produce colored broilers of both sexes, and lines to produce dark-skinned broilers. All of these developments by the secondary breeders will depend on the commitment they are willing to make to research (Hubbard, 1980).

A number of reproductive innovations are on the horizon. Artificial insemination will assume increasing importance in most breeders' programs. Recent breakthroughs in procedures for long-term freezing of chicken semen will allow breeders to greatly extend the use of outstanding sires. Frozen semen banks hold promise for extending the use of sires to the integrator level of the primary breeder's distribution program and to maintain lines that otherwise would be economically questionable. Artificial insemination may be extended to the parent stock if cage management and AI systems can be developed to reduce labor costs. If semen freezing techniques can be perfected so that frozen semen of high fertility (80+%) can be sold, the sale of breeder males could eventually decline in favor of the sale of frozen semen.

Dwarf broiler breeders will also assume increasing importance over the next few years. The dwarf breeder female is approximately 25% smaller than the standard type female, and even though her egg size is smaller and her progeny's growth rate is slightly less than that of the standard type broiler, the cost to produce the broiler chick from the dwarf breeder is less, and more than offsets the slight loss in her progeny's growth rate (for review, see Guillaume, 1976).

The field of biochemical genetics has been active in a number of poultry genetics laboratories. Discoveries such as genetic polymorphisms in proteins and enzymes (allozymes/isoenzymes), genetic analysis of blood groups that relate to production traits, the genetics and physiology of riboflavin metabolism, the hemoglobin system and possible mechanisms of regulatory gene control, and gene interaction in the control of melanin pigmentation have opened up completely new possibilities for studying the basic genetics of poultry breeding in greater detail (Sheldon, 1980). The technique of "biochemical breeding" or profiling is currently being used by one major primary chicken breeder, Hubbard Farms, Walpole, New Hampshire (Hubbard, 1980). Components of the blood (enzymes, hormones, lipoproteins, triglycerides) are being analyzed to identify gene markers for economic traits of importance such as growth rate, feed efficiency, fat deposition, and muscle production. Such a method will enable the breeders to reduce the time for selecting individual birds or families with special characteristics. This does not indicate that someday breeders will just select birds strictly on the biochemical profile but such biochemical parameters should be useful in enhancing genetic improvement. It could take 5–7 years before the merits of biochemical breeding reach the commercial producer.

One of the major breakthroughs made in poultry genetics in the past 25 years has been the discovery of genetic control of the immune response (Sheldon, 1980). Breeding for resistance to Marek's disease and lymphoid leukosis has become a standard strategy in the poultry industry as a result of the early studies on genetic control of the immune response. Although the advent of a Marek's disease vaccine during the last decade has reduced the incentive by commercial organizations to breed for resistance, many experts feel that vaccination and genetic resistance are required for adequate protection (Spencer *et al.*, 1974). Priority must be given to producing stock free from salmonellae, adenoviruses, mycoplasmas, and those Rheo viruses that cause viral arthritis. At present, programs in genetic resistance are in the hands of the research scientist, and the transition from the experimental laboratory to the commercial breeding company has yet to be achieved on a large scale basis. Future programs of disease control will be complemented by the establishment of programs in genetic resistance. Existence of genetically determined specific disease resistance has been established in both animals and plants. General resistance to disease has been demonstrated in varying degrees in plants, and field observations suggest that such general resistance might also exist in animals, but experimental evidence is not available (Gavora and Spencer, 1978). Observations in the field suggest that certain stocks of chickens have low mortality when raised in a number of different environments. It should also be noted that various pathogens develop resistance against drugs and antibiotics that are used for the control of specific diseases.

Although there is some bad news about the future developments in poultry breeding technology, fortunately the good news outweighs the bad. The modern advances, which have occurred in molecular genetics and in cellular techniques potentially useful in poultry breeding, have yet to be applied by the industry. There is some interest in exploitation of biochemical or physiological technologies (i.e., cloning, gene transfer, sex control) as practical approaches for poultry breeding (for review, see Shaklee, 1974), but progress toward commercial application is slow. Cloning, if such a technique could be developed, might be used to propagate the superior breeders, that is, to reproduce exact copies of a particular male and female in large numbers. This would have great potential for the primary breeder, but at present, the primary breeder's goal is to produce the model bird that someday might be a good candidate for cloning (Hubbard, 1980). The infusion of genes (gene transfer), either individuals or groups, into the host by biological vectors could be most useful to primary breeders. In birds with low production or silent gene products, the production of individuals with added quantities of a particular gene could be very beneficial. If the sex of the individual could be controlled at fertilization, the impact on efficiency of production by directing the hatch of males for broilers and females for table egg production could be enormous. However, this technology may be difficult to develop because in avian species, unlike mammals, the female determines the sex of the offspring (Hubbard, 1980). Another overall problem is that the economically important traits are controlled by several genes thus making the practical applications of the above techniques difficult. Also many of the poultry breeders feel that progress in the next 10 years will come from a better use of the know-how that already exists rather than from any breakthrough in such fields as biochemical genetics or genetic manipulation. New findings in the next 10 years can hardly be expected to influence commercial breeding schemes within the same decade.

A far less obvious effect on the development and application of new poultry breeding technologies concerns the fact that within the last 10–15 years there has been a considerable reduction of research effort in basic poultry genetics (Clayton, 1978; Sheldon, 1980). Sheldon reasons that the decline in research of basic poultry genetics is because of emphasis on the quantitative genetic approach that has influenced research structures and attitudes of the industry. As a result, the level of funding of basic research from which new techniques must arise has been reduced. An equally pronounced and alarming effect is the shortage of suitably qualified people for the poultry breeding industry. The number of commercial breeding companies worldwide probably does not exceed 30, and the declining trend seen in recent decades is primarily a reflection of economic circumstances. The drastic reduction of breeding companies severely limits employment prospects for qualified geneticists, which in turn has an inhibitory effect on student interest and growth of poultry science departments in universities.

VI. CONCLUSION

Considerable progress has been made in poultry breeding technologies in the past 30 years. Certainly, these developments in poultry genetics and reproductive biology are comparable to those for other domestic species. The development of future technologies will depend primarily on increasing the research activities in basic genetics/reproductive biology by improving the funding level. The potential benefits from new poultry breeding technologies will enable the industry to keep pace with the ever-growing demand for poultry products worldwide and at the same time to continue to provide consumers with a low-cost source of protein for the human diet.

REFERENCES

Clayton, G. A. (1978). *World's Poult. Sci. J.* **34,** 204-208.

Gavora, J. S., and Spencer, J. L. (1978). *World's Poult. Sci. J.* 34, 137-148.

Guillaume, J. (1976). *World's Poult. Sci. J.* **32,** 285-304.

Hubbard, W. (1980). *Broiler Industry.* (June issue).

Hutt, F. B. (1949). *In* "Genetics of the Fowl." McGraw-Hill, New York.

Kinney, T. B. (1974). *World's Poult. Sci. J.* **30,** 8-31.

Lake, P. E. (1978). *In* "Artificial Insemination in Poultry" (Ministry of Agriculture, Fisheries, and Food), Bull. 213. HM Stationary Office, London.

Lorenz, F. W. (1969). *In* "Reproduction in Domestic Animals" (H. H. Coles and P. T. Cupps, eds.). Academic Press, New York.

Marble, D. R., and Jeffrey, F. P. (1955). *In* "Commercial Poultry Production." Ronald Press, New York.

Sexton, T. J. (1979). *Beltsville Symp. Agric. Res.* **3,** 159-170.

Sexton, T. J. (1980). *Proc. 9th Int. Congr. Anim. Reprod. Artif. Insemin., Madrid* **2,** 527-533.

Shaklee, W. E. (1974). *World's Poult. Sci. J.* **30,** 256-278.

Sheldon, B. L. (1980). *World's Poult. Sci. J.* **36,** 143-173.

Spencer, J. L., Gavora, J. S., Grunder, A. A., Robertson, A. and Speckman, G. W. (1974). *Avian Dis.* **18,** 33.

U.S. Department of Agriculture (1977). *In* "Agricultural Statistics." U.S. Govt. Printing Office, Washington, D.C.

Van Wambeke, F. (1967). *J. Reprod. Fertil.* **13,** 571-575.

5

Fish and Aquatic Species

WALLIS H. CLARK, JR., AND ANN B. McGUIRE

I. INTRODUCTION

A. The Case for Aquaculture

Aquaculture—the cultivation of freshwater and marine species (the latter often termed *mariculture*)—is an ancient concept. China began culturing fish some 4000 years ago and probably has one of the highest numbers of aquacultural operations in any single country today. Even with this long history, however, the science of aquaculture is far behind that of terrestrial animal husbandry, but

91

NEW TECHNOLOGIES IN ANIMAL BREEDING

recent private and institutional research efforts are transforming the "art" of aquaculture into a modern, multidisciplinary technology.

This transformation has been fueled by the urgent need for increased food supplies and by the inability of the natural fisheries to supply their growing market. Fisheries products are well-recognized for their high nutritional quality and unique health benefits. They currently provide roughly 18% of the animal protein consumed in the world (Morris, 1980), and this figure is rapidly growing. Increasing demand for these products has added such pressure to the natural fisheries that the catch levels of many species, including herring, cod, haddock, flatfishes, shrimp, and lobster, have already reached, if not surpassed, their maximum sustainable yield (Sindermann, 1978). In fact, certain species (such as abalone, anchovy, and sardine) can no longer support commercial harvests in many traditional fisheries grounds.

This situation clearly exposes limitations to fisheries supplies. Traditional marine fisheries, once considered infinite, are now thought capable of producing a maximum sustained harvest of 100–120 million metric tons (MMT) annually (Glude, 1978).

As commercial fishing efforts expand and increase in efficiency, the harvests will approach this maximum and are expected to reach at least 94 MMT by the year 2000 (Glude, 1978). Such high fishing pressures combined with the needs of an expanding human population will create a demand for fisheries products that far outstrips the natural availability, resulting in higher prices and shortages.

Recognition of this increasing need for fisheries products has stimulated considerable interest in aquaculture on the part of national and international agencies, as well as private entrepreneurs. In addition to being sources of high-quality protein in great demand, many aquatic organisms are highly suited for culture because they devote more food energy to growth than terrestrial animals do (Bardach *et al.*, 1972). Surprisingly, many forms of aquaculture can produce more protein per unit of water than is produced through terrestrial agriculture, with the added benefit that the aquaculture wastewater can then be used to irrigate crops. With this knowledge, public institutions generally have concentrated their attention on the culture of protein-rich, low-cost aquaculture species low on the trophic level, whereas entrepreneurs have pursued species with high market demand that are high on the trophic level and require supplementary feeds. This interest has resulted in aquacultural research and development that run the gamut from primary species (blue-green and eucaryotic algae) to carnivores (lobsters and salmonids).

B. Aquaculture Production

As a result of this widespread interest, aquacultural ventures occur worldwide, with a variety of sites and production scales. For example, China grows mainly macroalgae (seaweeds) and carp; Japan cultures numerous marine organisms including yellowtail, sea bream, salmonids, tuna, penaeid shrimp, oysters, scallops, abalone, and algae; the USSR cultures fish such as salmon, sturgeon, and carp; North America produces catfish, trout, salmon, oysters, and shrimp; and Europe cultures flatfish, trout, oysters, mussels, and eels (Pillay, 1976).

In 1975, aquaculture constituted roughly 10–12% of the world's production of fish and shellfish, compared with an estimated 7% contribution in 1967 (Pillay, 1976). From 1970 to 1975, worldwide aquacultural production approximately doubled to 6 MMT. Most of the 1975 production—roughly 66%—consisted of freshwater, brackishwater, and marine fish; the remainder was made up of seaweeds (17.5%), molluscs (16.2%), and crustaceans (0.3%) (Pillay, 1976).

The 1975 aquacultural production in the United States totaled approximately 65,000 metric tons, and by 1978, it had increased to an estimated 100,000 metric tons (MT) (National Research Council, 1978). This figure is expected to reach 250,000 MT by 1985 and over 1 million MT by the year 2000 (National Research Council, 1978), as part of a worldwide surge in aquacultural production.

In the United States, where the industry is fairly new, commercial operations are currently profitable for the culture of catfish, trout, salmon, shrimp, crayfish, and oysters. Catfish is presently the most important species, with an estimated 1979 production of $42 million on an ex-farm basis (International Resource Development, Inc., 1980). Second in commercial importance is trout culture, which has been practiced in this country since the turn of the century and uses the most sophisticated culture systems available. Trout production was estimated at $22 million on an ex-farm basis for 1979 (International Resource Development, Inc., 1980). In 1979, salmon production, through cage culture and ocean ranching, amounted to an estimated $4 million (ex-farm basis), and the 1979 ex-farm production of shrimp, crayfish, and other crustaceans is estimated at $11 million (International Resource Development, Inc., 1980). Oyster culture is quite successful, using commercial hatcheries and low-intensity coastal growout methods, and together with production of other shellfish, oyster culture reached approximately $5 million (ex-farm basis) in 1979 (International Resource Development, Inc., 1980).

As mentioned, the aquaculture industry is expected to grow substantially during the 1980s. A recent report forecasts the value of shipments from United States aquaculture production to increase by 1989 to a total of $485 million (ex-farm, not considering inflation) compared with a production of roughly $89 million in 1979 (International Resource Development, Inc., 1980). Catfish production is expected to double by 1989 to approximately $83 million. Trout production, which may already have reached market demand, may increase slightly to approximately $36 million. Salmon culture should increase by more than 10-fold to $48 million in 10 years. The culture of shrimp, crayfish, and other crustaceans is expected to increase almost 8-fold to roughly $80 million, and the production of oysters and other shellfish is estimated to reach $140 million by 1989, an astonishing 28-fold increase (International Resource Development, Inc., 1980). These predictions do not take into account several species with strong aquaculture potential that have yet to enter the commercial production phase in the United States. These include striped bass, lobster, tilapia, carp, perch, clam, bass, eel, and sturgeon.

II. CURRENT TECHNIQUES OF CULTURE

A. Extensive Methods

Just as in terrestrial agriculture, the practice of aquaculture can range from low-technology, extensive methods to highly mechanized intensive systems. At one extreme, extensive aquaculture may simply be contained stock replenishment, where culturists enclose their animals in natural bodies of water (such as coastal embayments) and rely on the environment to supply the animals' requirements. Such culture entails minimal management and low investment and operating costs; consequently, yields per unit area are also low. In highly intensive aquaculture, on the other hand, the animals are grown in strictly regulated environments (in raceways or tanks) at high densities, and all their needs are supplied by the culturist. Although the initial investment and operating costs are high and continual management is necessary, intensive aquaculture results in high production levels. Many aquacultural operations fall somewhere between the extremes of intensive and extensive, using aspects of both according to their site, chosen species, and financial and legal constraints.

In the Mediterranean area, extensive culture methods are used successfully to grow various fish. Mullet, for example, are contained in large estuaries, where they are grown in polyculture with other fish. Mullet culturists stock their systems by trapping wild juveniles that voluntarily enter the estuaries; they rely on naturally occurring nutrients to supply their animals' food needs. Elaborate canals and sluices allow for water exchange. In such systems, marketable fish are typically harvested after 2–3 years of containment, and the production levels are relatively low (100–200 kg/ha) (Doroshov *et al.*, 1979).

Further up the scale of intensity are typical catfish farms in the Western Hemisphere. In such operations, catfish are stocked seasonally in large, freshwater ponds (up to 10 ha), and culturists supplement the ponds' natural food supply with dried, milled feeds. Within 2 years, the fish reach marketable size (0.7 kg), and production levels can be as high as 3 000 kg/ha (F. Conte, 1980, personal communication). Although such levels are far above those of extensive mullet culture, they are still limited by the culturists' lack of environmental control. For example, oxygen depletion can occur in densely stocked ponds and lead to massive fish kills. In an attempt to avoid such disasters, growers use various aeration devices, but with mixed results.

B. Intensive Methods

To increase production levels and limit risks, culturists (including catfish farmers) are turning to raceway and tank systems similar to those used for trout culture. One such enterprise in California rears catfish in tanks ranging in size from 45 to 135 m^3. Optimal water temperature is maintained year-round with geothermal water, and oxygen levels are carefully controlled. This system is capable of 1000-fold higher yields than in the pond culture of catfish, providing marketable fish (weighing 0.9 kg) within 1 year (R. Klosterman, 1980, personal communication).

Fig. 1. Experimental intensive system. Culturists are examining brood stock held in indoor raceway similar to those used in marine shrimp production systems. (Photo courtesy of Bernard Colvin, University of Arizona, Environmental Research Laboratory.)

Marine species are also being cultured intensively. The culture of penaeid shrimp (the common marine prawn) is still under development but appears close to commercial applicability. In these modern culture systems, the shrimp are raised in shallow raceways enclosed in inflatable polyethylene domes that resemble greenhouses. The animals are reared at high densities (3.3 kg/m^2), fed artificial feeds, and grown under strictly controlled environmental regimens (Fig. 1). The water either recirculates continuously or flows through the system with constant replacement. Recently, the penaeid shrimp's life cycle was closed in captivity, and hatchery technology for seed stock production has become available. With these advances, marketable animals can be produced within 4–6 months of growout (Doroshov *et al.*, 1979).

The two extremes of intensive and extensive aquaculture are uniquely combined in the ocean ranching of anadromous fish (fish that migrate from the ocean to freshwater for spawning). In these systems, young fish such as salmon and sturgeon are produced in highly intensive hatcheries and then released into their natural environment for growout. After reaching sexual maturity, these animals return to their site of birth, where they are harvested.

C. Research Directions in Intensive Aquaculture

Although extensive aquaculture is still practiced in many parts of the world, the more industrialized areas such as the United States and Europe are developing intensive culture methods, largely because of their limited land and water availability and the need for year-round, mass production of aquatic products. This technological shift is requiring considerable knowledge of the needs of each culture species and the ability to manipulate these animals throughout their life cycles. As a result, research programs are emphasizing the development of technology for reproductive control, hatchery systems, feed formulation, disease control, and systems engineering.

Reproductive control and subsequent genetic selection of aquatic animals is critical to the future of intensive aquaculture, as will be discussed in Section III. The elucidation of hatchery technology is also receiving research attention because the sensitivity and complex development of many aquatic animals at the embryonic and larval stages has led to high mortalities in intensive hatchery systems. As more biological information becomes available on the distinct feeding and environmental requirements of the early stages of aquatic animals, researchers are translating this knowledge into efficient hatchery techniques that allow high survival rates.

Increased emphasis is being placed on the nutrition of and feeds formulation for aquatic organisms of aquaculture potential. Currently, only a few aquatic species can be raised solely on prepared feeds; the remainder are fed expensive natural foods (such as brine shrimp and trash fish) either alone or as diet supplements. Before practical artificial foods can be formulated, however, nutritionists must define each species' nutrient requirements at all developmental stages and use that knowledge to devise inexpensive, formulated feeds that are water stable and palatable to the particular animal.

As in livestock production, disease control becomes important in high-density culture conditions in which pathogens can quickly wipe out entire stocks. Current research, therefore, includes the study of diseases of aquatic animals, as well as the formulation of prophylactic and therapeutic practices useful to the industry.

The trend toward controlled-environment aquaculture has also mandated increased sophistication of animal rearing systems. As a result, aquacultural engineers are currently emphasizing the design of practical systems and devices that are appropriate to all stages of each cultured species, are energy efficient, allow food delivery, and are suitable for maintaining good water quality.

Aquaculture will benefit not only from the increasing scientific attention previously described, but also from solutions to economic, legal, and social problems that currently limit growth of the industry. Particularly in industrialized nations, potential aquaculturists are often faced with numerous obstacles before they can begin culturing their chosen species. Some common problems include the possible illegality of introducing a nonindigenous species, resistance to procurement of wild animals for brood stock or seed stock, damage to cultured species by protected predators or competitors, and competing uses of public waters. Recently, however, governmental law- and policy-makers are becoming aware of these problems and seeking solutions. For example, in the United States, a National Aquaculture Plan has been adopted, and state and federal legislation, designed to encourage the industry, is being written.

III. APPLIED GENETICS: ANIMAL BREEDING IN AQUACULTURE

A. Introduction

Aquaculture is easily justified, considering the quality of protein provided by fishery products and the decline of the world's wild fisheries. Aquacultural practices, however, are hundreds of years behind those of terrestrial agriculture and, in most instances, are attempting to grow wild animals. Only a few cultured species are truly domesticated; these are carp, trout, and possibly tilapia. Though aquaculture needs assistance in most technological disciplines, the immediate need is in animal breeding. Far too often, the principal deterrent to a successful culture operation, whether extensive or intensive, is the unavailability of appropriate seed stock.

Part of this problem stems from the disparate body of knowledge concerning aquatic animal breeding. On the one hand, a wealth of information is available on such animals as the echinoderms (which include sea urchins and starfish), upon which we base much of our modern understanding of reproductive biology—particularly gamete biology. On the other hand, such knowledge is scant for the animals chosen for culture. Although some research has been done with aquatic animals on induction of gonadal development, gamete manipulation, chromosome manipulation, and selective breeding, this work is often fragmentary, and the results, for the most part, are not yet being applied commercially.

To discuss aquatic animal breeding in more detail, the following overview will be divided into three parts, corresponding to the major groups of aquaculture species: fish, molluscs, and crustaceans.

B. Reproductive Control in Fish

Certain fish, namely carp, tilapia, and trout, represent some of the most successful species used in aquaculture. Carp and tilapia have been cultured for thousands of years, and trout have been reared in the United States for almost a century. Each of these fish readily breeds in captivity, exhibits high fecundity and good larval survival, and thus, provides a relative abundance of seed stock.

With all three fish, selective breeding programs have long been established, healthy gene pools are available, and advantageous hybridizations have been developed (Moav, 1979). Interestingly, sterile hybrids are often desired to counteract early reproductive success in culture systems, particularly with carp and tilapia. These animals can attain reproductive maturation at an extremely small size, thereby overpopulating systems with unmarketable fish. Currently, techniques for the development of sterile stock through the induction of abnormal ploidy and hormone sterilization are being examined (Stanley, 1979). Such aquatic neuters would use their food energy to produce flesh rather than gonads.

Although highly sophisticated hatchery systems have been developed for salmon, selective breeding of this anadromous fish has been sharply limited in the United States because of political pressures. Since cultured salmon are used for ocean ranching and to replenish natural stocks, the concern has been that selected stocks could damage natural populations.

In the USSR, however, genetic work has proved highly successful in the culture of another anadromous fish—the sturgeon—a native animal for which intensive hatchery systems have been developed to replenish natural populations and supply seed for contained culture (Burtzev and Serebryakova, 1973). One result of this work has been the development of a sturgeon hybrid between *Huso huso* and *Acipenser ruthenus* that is well-suited for aquaculture systems (Romanycheva, 1977). Recently, research on sturgeon culture was initiated in the United States (McGuire, 1979). Although this project is young, scientists have hormonally induced ovulations in the white sturgeon, successfully fertilized the eggs, and raised the animals from the larval to the fingerling stages (McGuire, 1980). Unlike most farmed animals, which exhibit low fecundity, the sturgeon is a farmer's delight, often yielding 1 million eggs at ovulation (Fig 2).

Fig. 2. Typical ovulation collected from one white sturgeon after hormone induction.

Though the potential for sturgeon culture is exciting, the largest aquaculture industry in the United States currently is channel catfish farming. This fish is a newcomer to aquaculture, having been cultured seriously only since the mid-1960s. Because of this short history, a limited amount of work has been conducted on selective breeding in catfish, and the establishment of viable programs has been limited by financial constraints. Although fecundity is high in catfish and larval-rearing techniques are well-developed, the industry continues to suffer from a lack of reliable seed stock production. The problems arise from the unsophisticated brood stock and hatchery systems currently in use. In these systems, mature breeding fish are maintained in ponds, where, once each year, breeding pairs segregate, and the females lay their egg masses in containers provided (usually milk cans). The eggs are fertilized by the males, and these fertilized egg masses are retrieved from the containers and hatched in troughs, where they are continually aerated by mechanical paddles (Bardach *et al.*, 1972). Although reproductive research on catfish includes the use of chorionic gonadotropins to control maturation and induce spawning and the development of *in vitro* fertilization techniques (Clemens and Sneed, 1962), these methods are far from routine and are not yet being used by the catfish industry.

Unfortunately, many fish species suitable for aquaculture do not reproduce in captivity (e.g., striped bass, large mouth black bass, and pompano), including even some species historically cultured in extensive systems (e.g., mullet, milkfish, and eel) (Bardach *et al.*, 1972). In some instances, wild brood stock have been induced to mature and spawn using environmental manipulation, chorionic gonadotropin injections, or a combination of the two (National Research Council, 1978). Nevertheless, such methods still require capture of wild animals, and the techniques necessary to ensure a steady supply of seed stock are far short of commercial needs. In mullet and milkfish culture, systems are stocked with fry trapped from the wild, obviously precluding selective breeding.

A useful reproductive technique that has been under study with various fish species including salmon, trout, carp, and catfish is cryogenics. Thus far, freezing of gametes has been successful with sperm, but it appears that the technique cannot be applied to eggs (Horton and Ott, 1976). Because shipping and storing is easier with frozen sperm than with whole animals, the use of cryogenics in aquaculture undoubtedly will increase.

C. Reproductive Control in Molluscs

The second major category of aquaculture species is the molluscs, many of which are traditional foods throughout the world. Oyster culture, which began in Europe during Roman times, was probably the first practiced form of mariculture, and by 1950 breeding technology and hatchery systems were well-defined for oysters of the genus *Crassostrea* (Bardach *et al.*, 1972). These culture systems have now been applied to scallops and clams.

To prepare molluscs for spawning, reproductively mature animals are held at a certain maintenance temperature (which, for the American oyster, is approximately 7°–10°C). They are conditioned for spawning by slowly raising the temperature 5°–7°C and holding the animals at this new level for 2–4 weeks. At the end of this period, the temperature is quickly spiked an additional 10°C, and spawning ensues. The molluscs release their gametes, and external fertilization occurs (Bardach *et al.*, 1972). The resultant free-swimming larvae are then fed phytoplankton. Once the larvae become benthic, they can be used as seed stock for a variety of coastal growout methods.

Though the culture of another mollusc, the abalone, is still in the research and development stages, commercial viability appears imminent. To spawn abalone, culturists use heat shock methods as described above, exposure to ultraviolet light, or treatment with a dilute solution of H_2O_2 (Fig. 3) (Morse *et al.*, 1977,

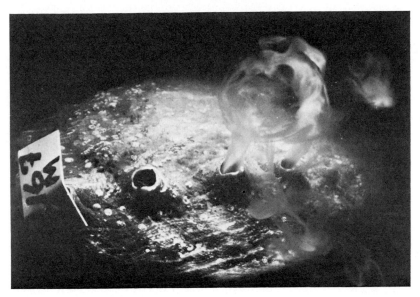

Fig. 3. Release of sperm from an abalone that had been induced to spawn with hydrogen peroxide. (Photo courtesy of Daniel E. Morse, University of California, Santa Barbara, Marine Science Institute.)

1978b). The latter technique is relatively new and, although it is used primarily with abalone, its effectiveness with other molluscs is being shown (Morse *et al.*, 1978a,b).

For the most part, molluscs are highly suited for selective breeding and other genetic manipulation because they are free spawners (i.e., they release their gametes directly into the water) and exhibit extremely high fecundity. The American oyster, which has a wide geographic range and genetic adaptations to numerous environmental conditions, appears to have sufficient genetic variance to respond well to selective breeding. This is evident in early work showing selection responses for growth rate and disease resistance (National Research Council, 1978). In addition, Japanese and American oysters have been crossed within and between species (National Research Council, 1978), but the hybrids have not been tested adequately, and commercial use of these crosses has not been considered.

Some limited genetic work has been done on abalone. Species crosses have been made in Japan (National Research Council, 1978), but genetic variability has not been measured, and selection experiments have not been conducted. Attempts at genetic improvement of growth are being initiated in an effort to counteract the abalone's slow growth, which is a major deterrent to its commercial production.

D. Reproductive Control in Crustaceans

The third major group of aquaculture animals is the crustaceans, many of which have held great fascination for aquaculturists. Unfortunately, the reproductive biology of these animals is the most atypical and poorly understood of the three categories.

Of this group, the crayfish represents the largest and most viable industry, with Louisiana being the world's major crayfish producer (Bardach *et al.*, 1972). The animals are either harvested as a secondary crop in conjunction with rice or raised as a primary crop on crayfish farms. Though hatchery techniques are available to complete the animal's life cycle in captivity, commercial culturists choose to rely completely on wild stock for brood seed. It is generally believed that a determination of the available stocks and a definition of their suitability for different culture environments could lead to improvements in the industry. Unfortunately, this is slow in coming.

Another crustacean commonly cultured is the Malaysian prawn. This is a freshwater shrimp that belongs to the genus *Macrobrachium* and is now farmed in many warmer areas of the world, including the southern United States and Hawaii. The animals breed in captivity, and hatchery systems have been well-defined, making seed stock readily available (Ling, 1969). The major drawback to Malaysian prawn culture is the shrimp's aggressive behavior, which necessitates low stocking densities in growout systems. Geneticists planning to breed these animals for docility, however, are faced with domesticated culture stocks developed from only two females (National Research Council, 1978); obviously, such stocks possess little variability. Geneticists, therefore, must crossbreed the domestic animals with wild stock that are often difficult to obtain. The potential use of such hybrids is strong, because the genus *Macrobrachium* includes over a dozen species, and new techniques for electroejaculation and artificial insemination are under development (Sandifer and Lynn, 1980). Also, experimental *in vitro* fertilization procedures recently have resulted in larval production of *Macrobrachium* and appear promising (P. Sandifer, 1981, personal communication). These techniques for controlling reproduction will circumvent the behavioral isolation that currently prevents mating between species and will, therefore, lead to hybrid formation.

Of all highly prized fishery items, the lobster certainly ranks among the top. Efforts to culture this animal have been made since at least the 1860s, and records of ill-fated attempts to commercially rear the animal can be found in the archives. On the other hand, hatcheries designed to replenish natural stocks have operated for years in North America and Europe (Bardach *et al.*, 1972). Until recently, these hatcheries relied heavily on the capture of berried (egg-carrying) females for seed stock, because reproduction was seldom achieved in contained systems. Further problems frustrating would-be culturists were the lobster's slow growth, aggressiveness, and low genetic variability.

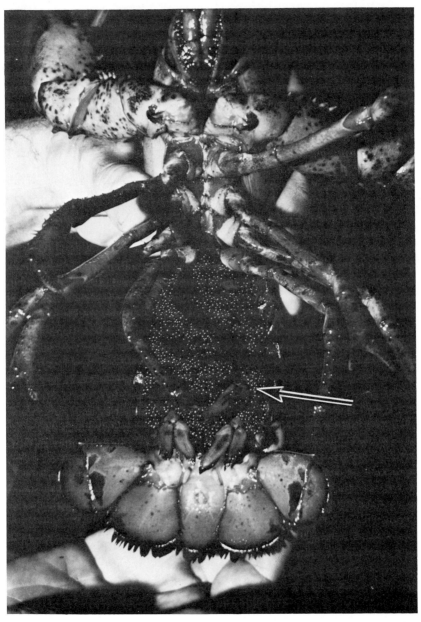

Fig. 4. A laboratory-reared lobster brooding a clutch of embryos (arrow).

This animal's continually increasing market value and its possibility of becoming an endangered species, however, maintained entrepreneurial and institutional interest in developing lobster culture techniques. As a result of this work, reproductive lobsters (Fig. 4) can now be maintained easily in captivity (Hedgecock *et al.*, 1978), and hatchery techniques are well-developed (Schurr *et al.*, 1976). In recent years, crosses between the North American lobster (*Homarus*

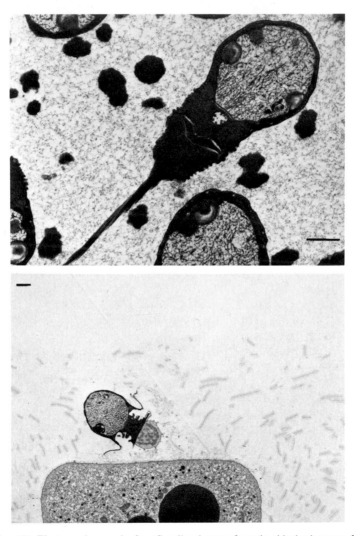

Fig. 5. (A) Electron micrograph of nonflagellated sperm from the ridgeback prawn. The single appendage, the "spike," represents the anterior end of the cell and is actually part of the sperm's acrosome. (B) Electron micrograph of an acrosome-reacted sperm at the egg surface. Bars = 1 μm

americanus) and the European lobster (*H. vulgarus*) have produced viable reproductive hybrids, providing geneticists with the variation necessary for selection (Hedgecock *et al.,* 1978). With these advances, prototype commercial systems are now being initiated.

The final crustacean to be discussed is the marine prawn (of the family Penaeidae), a highly prized animal that is the subject of the largest commercial fishery in the United States. Extensive commercial culture of these shrimp has been practiced for many years in the Far East (Bardach *et al.,* 1972), but attempts to transfer the technology to the United States have failed, largely because of the land and labor requirements. Through continued efforts by entrepreneurs and researchers in the United States, however, the technology for intensive marine shrimp culture has been vastly improved. Hatchery systems for rearing seed stock are sophisticated (Mock and Neal, 1974; Doroshov *et al.,* 1979), as are the growout techniques mentioned previously in this report.

Unfortunately, these developments have been hindered by dependence on wild brood stock. This dependence precipitated several research efforts designed to provide a better understanding of the reproductive biology of the Penaeidae. As a result, the reproductive cycle of these animals is now closed in captivity (Brown *et al.,* 1979), and techniques for the control and activation of these animals' atypical gametes (Fig. 5) are available (Clark and Lynn, 1977; Yudin *et al.,* 1979; Clark *et al.,* 1980). Also, *in vitro* fertilization methods for the marine prawn are now under development and appear imminent (Clark *et al.,* 1973). Availability of these techniques would lead rapidly to genetic manipulation of this animal, because the marine prawn is a free spawner unlike most crustaceans, which brood their eggs.

As discussed in this overview, aquaculture clearly suffers from an insufficient base of reproductive research on the species of interest. As this understanding is developed, culture technology and genetic manipulation of these animals will increase in sophistication, following the tradition of terrestrial animal husbandry.

REFERENCES

Bardach, J. E., Ryther, J. H., and McLarney, W. O. (1972). "Aquaculture: The Farming and Husbandry of Freshwater and Marine Organisms." Wiley, New York.
Brown, A., McVey, J., Middleditch, B. S., and Lawrence, A. L. (1979). *Proc. 10th Annu. Meet. World Maric. Soc.* **10,** 435–444.
Burtzev, I. A., and Serebryakova, E. V. (1973). *Genetics* **74,** 35.
Clark, W. H., Jr., and Lynn, J. W. (1977). *J. Exp. Zool.* **200,** 177–183.
Clark, W. H., Jr., Talbot, P., Neal, R. A., Mock, C. R., and Salser, B. R. (1973). *Mar. Biol. (Berlin)* **22,** 353–354.
Clark, W. H., Jr., Lynn, J. W., Yudin, A. I., and Persyn, H. O. (1980). *Biol. Bull. (Woods Hole, Mass.)* **158,** 175–186.

Clemens, H. P., and Sneed, K. E. (1962). "Bioassay and Use of Pituitary Materials to Spawn Warmwater Fishes," U.S. Bureau of Commercial Fisheries Research Report 61, Washington, D.C.

Doroshov, S. I., Conte, F. S., and Clark, W. H., Jr. (1979). *In* "The Biosaline Concept: An Approach to the Utilization of Underexploited Resources" (A. Hollaender, J. C. Aller, E. Epstein, A. San Pietro, and O. Zaborsky, eds.), pp. 261–283. Plenum, New York.

Glude, J. E. (1978). *In* "Drugs and Food from the Sea: Myth or Reality?" (P. N. Kaul and C. J. Sindermann, eds.), pp. 235–247. Univ. of Oklahoma Press, Norman.

Hedgecock, D., Moffett, B., Borgeson, W., and Nelson, K. (1978). *Proc. 9th Annu. Meet. World Maric. Soc.* **9,** 469–479.

Horton, H. F., and Ott, A. G. (1976). *J. Fish. Res. Board Can.* **33,** Pt. 2, 995–1000.

International Resource Development, Inc. (1980). *Aquaculture Magazine* pp. 4–9. (Buyer's Guide).

Ling, S. W. (1969). *FAO Fish. Rep.* **57,** 607–619.

McGuire, A. B. (1979). *Calif. Agric.* **33,** 4–6.

McGuire, A. B. (1980). *Aquaculture Magazine* **6,** 4–5.

Moav, R. (1979). *In* "Advances in Aquaculture" (T. V. R. Pillay and W. A. Dill, eds.), pp. 610–622. Fishing News Books, Ltd., Farnham, Surrey, England.

Mock, C., and Neal, R. A. (1974). "Penaeid Shrimp Hatchery Systems." FAO, United Nations. CARPAS 6-74-SE29.

Morris, J. G. (1980). *In* "Animal Agriculture" (H. H. Cole and W. N. Garrett, eds.), 2nd ed., pp. 21–45. Freeman, San Francisco, California.

Morse, D. E., Duncan, H., Hooker, N., and Morse, A. (1977). *Science (Washington, D.C.)* **196,** 298–300.

Morse, D. E., Duncan, H., Hooker, N., and Morse, A. (1978a). *FAO Fish. Rep.* **200,** 29–300.

Morse, D. E., Hooker, N., and Morse, A. (1978b). *Proc. Annu. Meet. World Maric. Soc.* **9,** 543–547.

National Research Council (1978). "Aquaculture in the United States—Constraints and Opportunities," 123 pp. Nat. Acad. Sci., Washington, D.C.

Pillay, T. V. R. (1976). *FAO Technical Conference on Aquaculture,* Kyoto 1975, FIR:AQ/Conf./76. Food and Agriculture Organization, Rome.

Romanycheva, O. D. (1977). *In* "Proc. 5th Japan-Soviet Joint Symp. on Aquaculture," Sept. 1976, pp. 353–366. Tokai University, Tokyo.

Sandifer, P. A., and Lynn, J. W. (1981). *In* "Advances in Invertebrate Reproduction" (W. H. Clark, Jr., and T. S. Adams, eds.), pp. 271–288. Elsevier, New York.

Schuur, A. M., Fisher, W. S., Van Olst, J., Carlberg, J., Hughes, J. T., Schleser, R. A., and Ford, R. (1976). "Hatchery Methods for the Production of Juvenile Lobster (*Homarus americanus*)." UC Sea Grant College Program, SG pub. No. 48. San Diego, California.

Sindermann, C. J. (1978). *In* "Drugs and Food From the Sea: Myth or Reality?" (P. N. Kaul and C. J. Sindermann, eds.), pp. 233–234. Univ. of Oklahoma Press, Norman.

Stanley, J. G. (1981). Presented at *Second ICES/ELH Symposium,* "Early Life of Fish." Woods Hole, Massachusetts (in press).

Yudin, A. I., Clark, W. H., Jr., and Kleve, M. G. (1979). *J. Exp. Zool.* **210,** 569–574.

III

Developing
Technologies

6

Approaches to Sex Selection in Farm Animals

**KEITH J. BETTERIDGE, W. C. D. HARE, AND
ELIZABETH L. SINGH**

I. INTRODUCTION

Man's desire to control the sex of his animals must be almost as old as his domestication of them. Certainly there is a lively folklore on the topic which

109

NEW TECHNOLOGIES IN ANIMAL BREEDING
Copyright © 1981 by Academic Press, Inc.
All rights of reproduction in any form reserved.
ISBN 0-12-123450-9

extends back to ancient times and is still being perpetuated. Smith (1919) has chronicled many of the older ideas on the subject: In 456 B.C., it was thought that females came from the left testicle and males from the right; Pliny knew in 50 A.D. that bulls dismounted to the right after siring a male and to the left after begetting a female; the Geoponica (a work that also revealed that the double hump of the Bactrian camel was due to impregnation by a wild boar) stated that sex was determined by wind direction at the time of copulation; and George Turberville, writing in sixteenth century England, was adamant that, for breeding hounds, sexual intercourse under the sign of Gemini and Aquarius, after the moon has passed the full, produced mainly males, and males ''less prone to madness'' at that.

Similar tales find surprisingly avid acceptance even now. Another Englishman's ''success'' in altering the sex ratio of calves by appropriate north-south or east-west orientation of his cows at the time of insemination, for example, recently enjoyed worldwide publicity. Scientific questioning of the claim was popularly reported as the ''sour grapes'' reaction of those who had been unable to achieve as much. Not infrequently, though, there are more serious difficulties to be faced in assessing the validity of the latest scientific claim that a means of sex control has at last been found.

In this chapter, we shall classify the approaches that currently seem possible and offer our assessment of their relative feasibilities. The classification is based very largely on a recent review (Hare and Betteridge, 1978, to which the reader is referred for a more complete bibliography of work up to that date), but is supplemented with new data that have come to light over the 2 years since then.

II. GENERAL PRINCIPLES OF PRENATAL SEX DETERMINATION

The genetic sex of a mammal is set at fertilization and depends on whether the X-bearing haploid ovum is fertilized by an X- or Y-bearing haploid spermatozoon. Thus, sex could be *predetermined* if X and Y spermatozoa were separated before insemination, and this would undoubtedly be the ideal method of sex selection. Sex would also be predetermined in embryos resulting from transplantation of diploid nuclei into enucleated ova, parthenogenetic activation of ova, development of a fertilized ovum after removal of one pronucleus, or the fusion of two oocytes.

The established genetic sex of a mammalian embryo or fetus can be *diagnosed* by examining diploid cells at a suitable stage after fertilization to see whether their sex chromosomes are XX (female) or XY (male). This can be done by examining cells for sex chromatin or a Y body, by chromosomal analysis, or, possibly, by the detection of an antigen that is peculiar to cells containing a Y chromosome. Cells for analysis can be obtained from embryos in the course of embryo transfer or from fluids aspirated from conceptuses at later stages of gestation.

Sexual differentiation of an embryo into a male or female depends not only on its genotype but also on the hormonal environment in which it develops. Normally, the hormones are of fetal origin, and, since they differ to some degree between the sexes, measurement of hormone levels in fetal fluids might also be useful for diagnosing gonadal sex prenatally. Experimentally, injection of androgens into genetically female fetuses masculinizes them to a variable extent. It should be emphasized, though, that this is only a phenotypic modification and that no functional sex reversal has ever been recorded in any mammal (Short, 1979).*

Sex predetermination by separating spermatozoa would control the sex of all embryos and the selection of sexed embryos before transfer limits pregnancies to those bearing the required sex. However, selection of sexed fetuses at later stages involves making alternative use of dams carrying fetuses of the unwanted sex. If they are to be aborted, the diagnostic method has to be practicable early in gestation, but even methods limited to use late in pregnancy could be helpful under some management conditions. The advantages and disadvantages of available and potential methods of prenatal sex control or selection, therefore, must be considered in relation to the use that is to be made of the procedure.

III. METHODS OF SEX PREDETERMINATION

A. Sperm Separation

Approaches to achieving this ideal can be divided into attempts to separate and attempts to differentially inactivate X- and Y-bearing spermatozoa (for reviews, see Beatty, 1974; Hare and Betteridge, 1978; Short, 1979).

Separation methods have been based on presumed differences between haploid spermatozoa expressed as mass, electrical charge, or motility.

Differential inactivation attempts have involved the treatment of semen with hormones, chemicals, altered atmospheric pressure, or antibodies to Y-bearing spermatozoa, as well as immunization of the female against H-Y antigen.

To quote from Short (1979):

Many workers have claimed to have separated X- from Y-bearing spermatozoa by a variety of techniques ranging from the ingenious to the incredible, and each year that passes seems to

---

*A 1980 press release from a reputable university, headed "Researchers Change Female Sheep Fetuses into Males," suggests that producers are given "... the option of having all females in any given year or a female from any given female in the herd...." This release is misleading, to say the least, and a cautionary example of what can result from a publicity agent's desire to satisfy the public's appetite.

bring a new claimant. The fact that nobody has been able to confirm any of these claims suggests that, for the moment, this separation must be ranked amongst Nature's impossibilities.

Although we concur with this pessimism, we recognize that the search for separation methods will continue because of the potential rewards of a method that would reliably change the sex ratio, however marginally. It remains relevant, then, to consider the means of assessing how effective any particular separation method has been. Until recently, this has called for insemination trials to determine the sex ratio of progeny. These must be prolonged and, therefore, expensive to be statistically reliable. However, in man, a phenotypic difference in mass between X- and Y-bearing spermatozoa has now been proven (Sumner and Robinson, 1976). Moreover, when human spermatozoa are stained with quinacrine dihydrochloride, the distal end of the long arm of the Y chromosome fluoresces brightly and shows up as a bright dot, the F or Y body (Pearson, 1972), allowing the Y-bearing spermatozoon to be recognized. Thus, in man, the success of separation attempts can now be evaluated without resorting to progeny sex ratio determinations. Indirectly, this may help animal work, because separation methods that could be shown to work for the human being might also be applied to other species.

Unfortunately, it is unlikely that the same technique can be directly applied in domestic animals, because it is generally agreed that the Y chromosome in these species does not fluoresce (Pearson *et al.,* 1971; Ericsson *et al.,* 1973), although there are persistent claims to the contrary (Bhattacharya *et al.,* 1976, 1977). Since Bhattacharya and his colleagues are using a controversial test as the main criterion of success for their separation methods, progeny sex ratios should still be regarded as essential for corroboration of any claim to have separated X- and Y-bearing spermatozoa in the domestic species.

If necessary, the determination of such progeny sex ratios might be expedited in the future by using the semen fractions to inseminate superovulated females (or for *in vitro* fertilization) followed by embryo sexing as described later.

The remaining four approaches to predetermination of sex would also involve embryo manipulation of various sorts.

B. Nuclear Transplantation

"Phenocopies" produced in this way are of the same sex as the donor of the nuclei. This technology is at an advanced stage of development in amphibia, as described in Chapter 9, this volume.

Ova are very much smaller in mammals than in amphibia and are, therefore, more difficult to work with. However, the prospect of combining analogous techniques with embryo transfer to control sex in farm animals has been improved a little by two recent developments. First, there has been limited success in the experimental introduction of nuclei from rabbit embryonic cells into unfertilized rabbit eggs by microinjection or virus-induced fusion (Bromhall, 1975). Second, Modlinski (1978) has described the development of tetraploid mouse blastocysts following transfer of nuclei from early embryos into fertilized 1-cell eggs in 4 of 166 attempts. More work is essential before the feasibility of using this approach to sex control can be assessed.

C. Parthenogenesis and Removal of a Pronucleus from Fertilized Eggs

The topics are dealt with in Chapter 10, this volume. In the context of sex control, parthenogenesis is still not a practical means of producing solely female offspring because, even in laboratory animals, embryos induced to develop from eggs in this way do not progress beyond the forelimb-bud stage.

Embryos remaining viable after removal of a pronucleus and subsequent treatment with cytochalasin B are also bound to be female. Early work in mice produced live young, which were later shown to be fertile (Hoppe and Illmensee, 1977; Markert and Petters, 1978), so it is a matter of considerable disappointment that more recent efforts have been unsuccessful (see Chapter 10, this volume). Even if this setback is temporary, it must be appreciated that future application of analogous techniques in farm animals might be limited by recessive lethals causing embryo wastage.

D. Oocyte Fusion

The possibility of initiating embryo development by fusing two oocytes has been demonstrated (Soupart et al., 1978) and is worthy of further investigation as a means of producing females only.

IV. METHODS OF PRENATAL SEX DIAGNOSIS

A. Sex Chromatin and/or Y Body Identification

In some species, this provides a rapid and useful approach to determining the genetic sex of embryos at transfer and of fetuses later in gestation.

The sex chromatin method depends upon the identification of a dark staining body, approximately 0.8×1.1 μm in size, lying adjacent to the nuclear membrane in fixed and stained interphase cells. The body, known as a Barr body, is seen in a proportion of female cells and represents the heterochromatic, inactive, X chromosome. In keeping with the fact that there is generally only one active X chromosome, individuals have as many Barr bodies as they have X chromosomes in excess of 1.

The Y body method is based on the fact, previously mentioned, that part of the human Y chromosome fluoresces brightly when cells have been stained with quinacrine mustard or dihydrochloride and shows up in interphase cells as a bright dot.

It was by sex chromatin identification that Gardner and Edwards (1968) achieved the first successful sexing of embryos at the time of transfer. For the analysis, they removed 200–300 trophoblast cells from 5¾-day rabbit blastocysts by microsurgery.

Both sex chromatin and Y body identification have been used in various approaches to determining fetal genetic sex in man. These include: detection of the Barr body or fluorescing Y chromosome in uncultured amniotic fluid cells, detection of a fluorescing Y chromosome in chorionic cells found in cervical mucus, and detection of a fluorescing Y chromosome in fetal cells in the maternal blood stream (for references, see Hare and Betteridge, 1978).

In cattle, sheep, goats, pigs, and horses, however, sex chromatin detection is impractical because of the coarse granular nature of the chromatin, and Y body identification does not work because, as mentioned earlier (Sec. III, A), the Y chromosome does not fluoresce.

A limitation of both sex chromatin and Y body identification is that, unless they can be combined, as in man, they fail to identify sex chromosomal abnormalities such as XXY.

B. Sex Chromosomal Analysis

This is done on chromosomes from cells in metaphase. If sufficient cells are in mitosis, chromosome preparations can be made by a direct method; if not in mitosis, the cells have to be cultured and harvested at the appropriate time.

Examination of metaphase spreads prepared directly from cells removed from hatched embryos (i.e., free of zonae pellucidae) has been used to sex both bovine and ovine embryos at transfer (for references, see Hare et al., 1978). The procedures involved in cattle are summarized in Scheme 1. Their complexity means that a team of two cytogeneticists cannot be expected to analyze more than 12–15 embryos per day.

Personal communication of results from four laboratories plus our own findings indicate that 68% of 353 day-12 to -15 bovine embryos have been sexed successfully, but that only 33% of 110 sexed embryos transferred have resulted in pregnancies (Table I). An unidentified "donor factor" appears to influence results. It has not yet been possible to preserve embryos of this developmental stage by freezing. This constitutes a considerable limitation to the method, because it places restrictions on the number of embryos that can be sexed in the time that can be allowed between their collection and transfer.

Direct chromosomal analysis has also been used on cells removed from unhatched, days 6 and 7 cattle embryos at transfer with subsequent birth of sexed

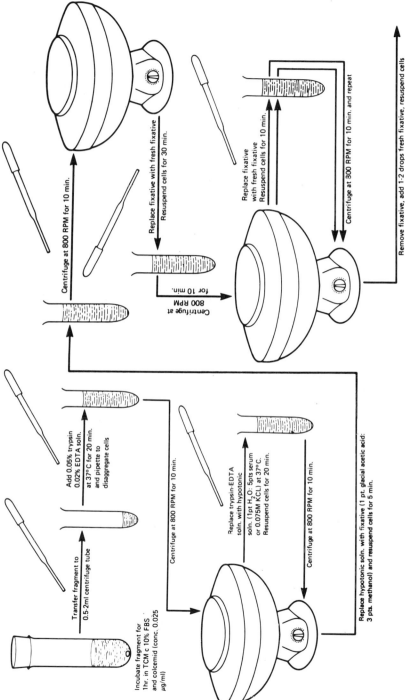

Incubate fragment for 1hr. in TCM c 10% FBS and colcemid (conc. 0.025 μg/ml)

Transfer fragment to 0.5–2ml centrifuge tube

Add 0.05% trypsin 0.02% EDTA soln. at 37°C for 20 min. and pipette to disaggregate cells

Centrifuge at 800 RPM for 10 min.

Replace trypsin-EDTA soln. with hypotonic soln. (1pt H₂O: 5pts serum or 0.075M KCL) at 37°C. Resuspend cells for 20 min.

Centrifuge at 800 RPM for 10 min.

Replace hypotonic soln. with fixative (1 pt. glacial acetic acid: 3 pts. methanol) and resuspend cells for 5 min.

Centrifuge at 800 RPM for 10 min.

Replace fixative with fresh fixative Resuspend cells for 30 min.

Centrifuge at 800 RPM for 10 min.

Replace fixative with fresh fixative Resuspend cells for 10 min.

Centrifuge at 800 RPM for 10 min. and repeat

Remove fixative, add 1–2 drops fresh fixative, resuspend cells

SCHEME 1

(*continued*)

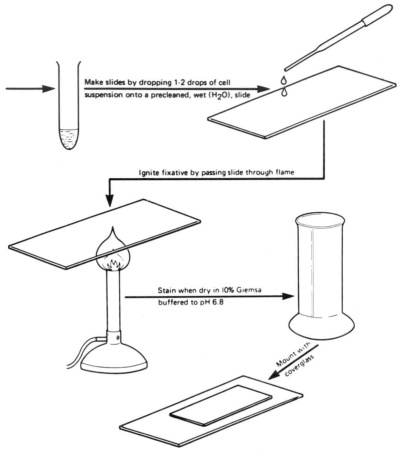

Make slides by dropping 1-2 drops of cell suspension onto a precleaned, wet (H$_2$O), slide

Ignite fixative by passing slide through flame

Stain when dry in 10% Giemsa buffered to pH 6.8

Mount with coverglass

SCHEME 1—(*Continued*)

calves (Moustafa *et al.*, 1978; Schneider and Hahn, 1979). An overall view of the laboratory apparatus required is seen in Fig. 1 and some steps in the procedures are shown in Figs. 2–5 and in Scheme 2. The overall sexing rate (59%) was similar to that achieved with hatched embryos, but the known tolerance of freeze preservation by the younger embryos would seem to make them preferable for eventual provision of frozen sexed embryos. However, in this laboratory (Singh and Hare, 1980), only 33% of embryos could be sexed in this way, largely due to a much lower average mitotic index in the embryos examined than observed in embryos studied by Moustafa *et al.* (Table II). In the authors' series, a donor factor again seemed to be involved. It is important that the reasons for these discrepancies be identified.

TABLE I

Results of Sexing and Transferring 12- to 15-Day-Old Embryos

	Embryos				
Number	Age (days)	Number sexed (%)	Number of sexed embryos transferred	Number of pregnancies (%)	Laboratory[a]
6	12	4 (66.6)	—	—	D
26	13	15 (57.7)	7	1 (14.3)	D
117	14	87 (74.3)	4	2 (50)	B
40	14	26 (65)	6	2 (33.3)	C
31	14	23 (74.2)	25	10 (40)	D
69	12–15	41 (59.4)	29	12 (41.4)	A
21	12–15	20 (95.2)	20	3 (15)	E
43	15	25 (58.1)	19	6 (31.6)	D
Total 353	12–15	241 (68.3)	110	36 (32.7)	

[a] Coded for anonymity

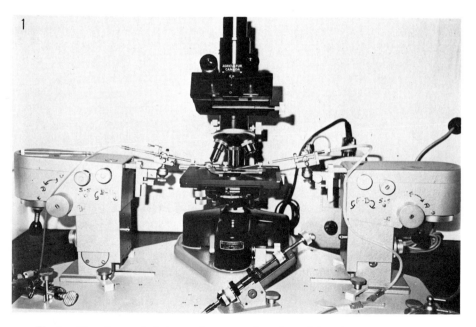

Fig. 1. The micromanipulation equipment consists of two Leitz micromanipulators with microinstruments, which are used in conjunction with a Laborlux microscope. The microinstruments, holding and suction pipettes, seen in Figs. 2–5 are made using a vertical pipette puller and a microforge.

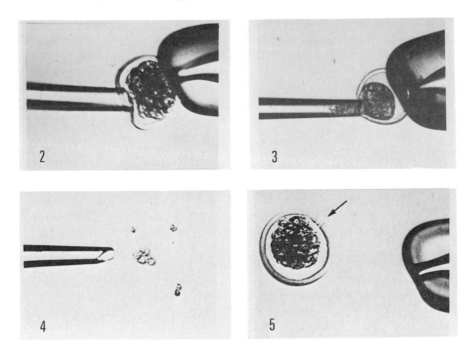

Fig. 2. Embryo is anchored during microsurgery by the holding pipette shown on the right. The zona pellucida is then punctured by the suction micropipette.

Fig. 3. Cells being aspirated into the suction micropipette.

Fig. 4. Cells that were removed from the embryo.

Fig. 5. Embryo after microsurgery. The arrow indicates the point of entry of the suction micropipette.

TABLE II

Results of Sexing 6- to 7-Day-Old Embryos

Embryos		Cells aspirated (number)	Preparations with >1 metaphase (%)	Embryos manipulated that were sexed (%)	References
Number	Age (days)				
44	6–7	7–10	77	59	Moustafa *et al.* (1978)
18	6	15–17	61	33	Singh and Hare (1980)

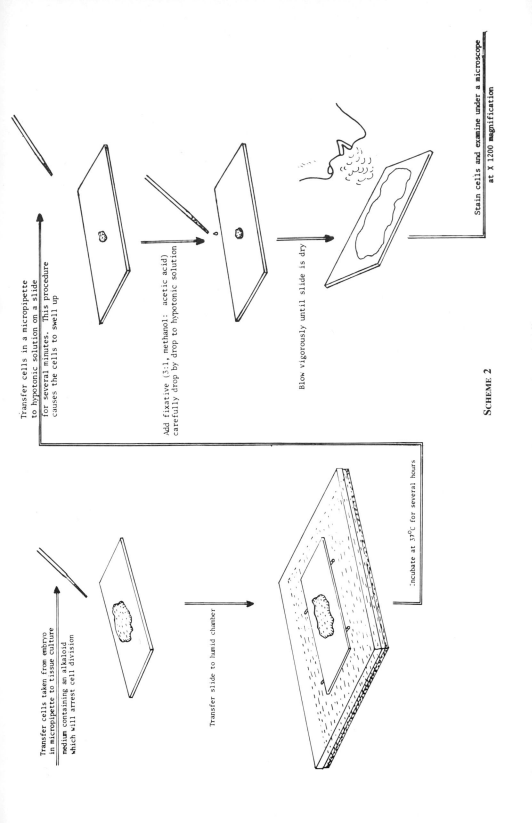

Transfer cells in a micropipette to hypotonic solution on a slide for several minutes. This procedure causes the cells to swell up

Add fixative (3:1, methanol: acetic acid) carefully drop by drop to hypotonic solution

Blow vigorously until slide is dry

Stain cells and examine under a microscope at X 1200 magnification

Transfer cells taken from embryo in micropipette to tissue culture medium containing an alkaloid which will arrest cell division

Transfer slide to humid chamber

Incubate at 37°C for several hours

SCHEME 2

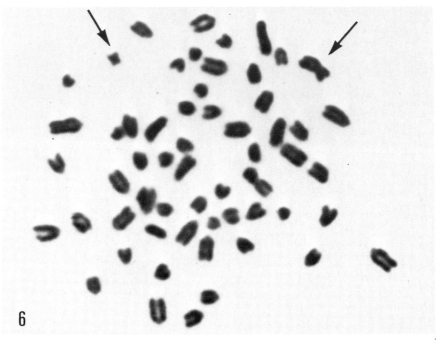

Fig. 6. A metaphase spread of bull chromosomes. Arrow on the left indicates the Y chromosome; arrow on the right, the X chromosome.

Cattle fetuses can be sexed by chromosomal analysis of cultured fetal fluid cells obtained by amniocentesis between days 70 and 90 of pregnancy,* which allows time for dams with calves of unwanted sex to be safely aborted (Eaglesome and Mitchell, 1977; Singh and Hare, 1977).

Cattle work has been facilitated to some extent by the fact that the bovine sex chromosomes are readily distinguishable from the autosomes in cells in metaphase during mitotic division (Fig. 6). This difference is also true of the dog but, in other farm species, the X chromosomes resemble some of the autosomes and so sex determination depends on having a full complement of chromosomes in the spread and being certain of the presence or absence of the Y chromosome.

C. Demonstration of H-Y (Male) Antigen

A possible alternative method of sexing embryos and fetuses is the serological demonstration of H-Y antigen. H-Y antigen is a histocompatibility antigen found on the surface of male, but not female, cells. It was recognized originally in a

---

*The term "amniocentesis" is used here to indicate the technique used, but it is not possible to distinguish between amniotic and allantoic fluids at the time of collection.

closely inbred strain of mice when the females rejected male, but not female, skin grafts. An antiserum to H-Y antigen can be produced in the same inbred strain of mice by inoculating females with male cells over a period of time. Using this antiserum, it has been shown that H-Y antigen is common to several species, including rats, rabbits, guinea pigs, human beings, cattle, and dogs.

Detection of H-Y antigen has been based on using the cells under test to absorb mouse anti-H-Y-sera and then measuring the reduction in activity of the sera in sperm cytotoxicity tests or various hemadsorption tests (see Wachtel, 1979). However, these tests are unsuitable for detecting H-Y antigen in early embryos which are to be transferred. Short (1979) cites new work by Nagai *et al.* and Ohno *et al.* on the purification of H-Y antigen from tumor cell lines, which may lead to more specific antisera and improved immunological means of detecting H-Y antigen bearing cells. Attempts have been made to develop an indirect immunofluorescent assay for H-Y antigen, but nonspecific staining and other technical problems have made interpretation of the results difficult (E. L. Singh and W. C. D. Hare, unpublished observations). To date, the assay has not been sufficiently developed to be used as a method of sexing embryos.

If H-Y antigen detection should develop into a sexing tool, a limitation of the method will be that, like sex chromatin or Y body identification, it will be subject to misinterpretation in certain sex chromosomal disorders. For example, on the one hand, it will not detect XXY males; on the other, some human, caprine, and canine XX male pseudohermaphrodites have been shown to be H-Y antigen positive.

D. Hormonal Assays

The possibility exists (for references, see Hare and Betteridge, 1978) of diagnosing sex prenatally by assay of fetal fluids for hormones (particularly those produced by the testis). This could be considered as an alternative to chromosomal analysis of cells from the same material collected by amniocentesis or laparoscopy.

To be useful for sexing, a hormone must (a) be measurable, (b) exist in fluids or tissues obtainable from the conceptus, and (c) differ between the sexes unambiguously. So far, the only group of hormones to have met most of the above criteria are fetal testicular androgens. The fetal testis becomes active by about 25–30 days of gestation in pigs, by 30 days in sheep, and by 45 ± 3 days in cattle. Its secretions begin to control differentiation of the male genital tract from these early stages. Although the androgens are peculiar to the male gonad at this time, distinction between males and females becomes less clear-cut than might be expected when the androgens are assayed in blood or fetal fluids. This is because androgens are found in both sexes due to steroid interconversions in the fetoplacental unit. Nevertheless, testosterone and androstenedione concentrations in the blood of the fetal calves are significantly higher in males than in females, decreasing progressively in both sexes from 90 to 260 days of gestation. Similar patterns are reflected in fetal fluids. Testosterone concentrations in allantoic fluid, obtained at slaughter, have been used to predict fetal sex in cattle between 90 and 150 days of gestation. Pure allantoic samples are necessary because sex differences between testosterone concentrations in amniotic fluid are less distinct. Since it is impossible to distinguish between the appearance of amniotic and allantoic fluids at collection from living animals at 70–90 days of pregnancy, the purity of the sample would have to be verified by the morphology and staining characteristics of the cells within it.

In man, there is disagreement as to whether the higher androgen levels in amniotic fluid of male fetuses can be used diagnostically. Some workers have used them to determine sex before 20 weeks of pregnancy and claim that the method offers advantages of speed and economy over chromosomal analysis. Others, however, have found the method inaccurate. Differences in assay sensitivity and specificity probably account for this discrepancy.

The extent to which androgens cross into the maternal circulation varies with species. Heifers carrying male fetuses have higher concentrations of testosterone in peripheral blood than do those carrying females. In rhesus monkeys, too, group comparisons show that androgen levels are significantly higher in maternal blood when the fetus is male. However, in both species, the differences are too small, and there is too much overlap between groups for androgen analysis to be used on individuals for sex determination.

Another hormone that is evidently peculiar to the testis in fetuses is the Mullerian inhibiting hormone, which inhibits female organogenesis. However, it is not yet known whether it can be detected in fetal fluids and could therefore be useful in sex diagnosis.

Fetal ovaries produce estrogens transiently near the time of sexual differentiation, and sulfated estrogens can be detected in the expanded ovine chorionic sac from as early as day 31 of pregnancy in sheep and from day 32 in cattle. However, these estrogens show no prospect of being useful for distinguishing between the sexes.

Estrone levels in bovine fetal blood are similar in males and females at 90–180 days gestation, but higher in males than females at 260 days. Estrone levels are also higher in maternal uterine venous blood for cows with male fetuses.

There is no sex difference in progesterone concentrations in fetal blood in cattle or sheep, although female rhesus monkey fetuses have higher levels than males.

In horses, fetal gonads of both sexes produce androgenic steroids that are aromatized in the placenta to estrogens detectable in the pregnant mare's blood and urine between about 150 and 250 days of gestation, but it is not known whether fetal sex influences the levels or nature of these steroids.

As to placental hormones, no relationship between levels of pregnant mare serum gonadotrophin (PMSG), nor of the ovine or bovine placental lactogens and fetal sex has been reported.

The use of hormonal assays for sexing embryos during transfer is probably remote, but estrogens are produced as early as day 12 (well before gonadal differentiation) in pig embryos. No influence of sex on their production has been reported, and, in any case, it is not yet possible to transfer pig embryos beyond day 9. Similarly, the preattachment horse embryo produces estrogens and androgens as early as day 10, but it is not known whether steroid production varies with sex or whether blastocysts large enough to provide an aliquot of fluid for analysis could survive transfer.

There is some recent evidence that weight differences between male and female rats become apparent *in utero* before sexual differentiation, presumably reflecting earlier sex-linked growth control. This raises the question of whether substances other than the hormones considered above might some day be of use in sexing.

V. DISCUSSION

There is a circuitous relationship between the development and application of prenatal sex determination techniques: Demand depends upon the cost and practicability of the technique; refinement of methods depends upon the emphasis placed on developmental work by the animal production industry. There is no doubt that an ideal method of sex predetermination, such as separation of X- and Y-bearing spermatozoa, would find widespread acceptance and use. Since this is not an immediate prospect, we shall focus our discussion on the methods of prenatal diagnosis of sex that are already available.

There are two components to diagnostic approaches that need to be considered: the actual diagnostic procedure (sex chromatin identification, chromosomal analysis, H-Y antigen detection, hormonal assay) and the material to which it is applied (from fetuses *in utero* or from embryos during transfer).

Techniques for obtaining fluids and cells from the fetus seem unlikely to become very much more efficient than they are already in cattle, but their possible extension into other species is largely unexplored. The sexing of the material obtained might well be improved (by detection of H-Y antigen, for example), but the limiting factors seem to be already set by the difficulties of sampling and the necessity of making alternative use of females carrying fetuses of the ''wrong'' sex. Essentially, then, sexing by amniocentesis can be seen as a developed technique ready for use if the demand is there. Demand seems likely to be restricted to research and highly specialized production applications.

Techniques for obtaining cells (and/or fluids) from embryos during transfer, on the other hand, can be extremely sophisticated if warranted. Since the technology of embryo transfer is also advancing rapidly, this seems to be a more likely avenue of progress to sex diagnosis and one that is by no means fully developed.

Even the question of the best developmental stage of embryo to use has not yet been resolved. Until noninvasive diagnostic methods (e.g., immunofluorescent detection of H-Y antigen) become practicable, an embryo has to be biopsied if it is to be sexed. In cattle, hatched embryos offer advantages as far as ease of biopsy is concerned and could become much more useful if storage techniques (notably freezing) can be found, thereby allowing more time for analysis of the cells obtained. Unhatched embryos at day 6 or 7 offer the advantage of established "freezability" and of being the best-understood stage in terms of nonsurgical collection, assessment, and transfer. However, the important discrepancies that have become apparent between different groups' success rates in sexing them by chromosomal analysis must be resolved before they can unreservedly be described as the stage upon which future efforts to improve sexing techniques must be concentrated. Very early cleavage stages (2–4 cell) have now been shown to tolerate microsurgical subdivision into blastomeres capable of separate development into identical live-born lambs and foals (Willadsen, 1979; Willadsen et al., 1980). Blastocysts developing from separated blastomeres can also be successfully preserved by freezing (Willadsen, 1980) and could therefore be held while their sex is being established by diagnostic procedures applied to the progeny of a companion blastomere. The diagnosis could be made by any convenient method at any convenient stage, even at birth! This approach, although demanding microsurgical facilities, seems quite promising in cattle now that advanced, nonsurgically collected embryos can at least be divided into identical twins (Willadsen et al., 1981).

In vitro fertilization might also bring the prospect of sex predetermination by oocyte fusion a little closer, and so it should be reiterated that all the predetermination possibilities would also ultimately depend on (or, in the case of separation of spermatozoa, be helped by) embryo transfer techniques.

If the development of sex control seems likely to owe much to concurrent developments in the embryo transfer industry, it also has much to offer to that industry in exchange. The export of Canadian Holstein embryos to Europe, for example, would be helped considerably if only females were sent, because the males are not required, and the supply of suitable recipients poses a real practical problem in many markets. Guaranteed females might also expand the market because Europe's emphasis on dual-purpose breeds is based on the need to produce beef from its dairy herd by using breeds whose male calves will fatten (like the Friesian). A European producer who knew he was buying only replacement heifer calves might look for a more specialized dairy breed (such as the Canadian Holstein). The fact that calving problems are fewer with female calves than with males (Burfening et al., 1978) would be an additional advantage of buying female embryos selectively.

Another repercussion can be foreseen if the dairy industry can be provided with guaranteed female embryos at a reasonable cost. Heifers would be used to carry female embryos from proven donors and would thereby be removed from the population that is presently artificially inseminated. This could reduce the AI industry's revenue and bring about its own involvement in embryo transfer for purely economic reasons. Infusion of that kind of support could go a long way to completion of the circle by improving sexing techniques and widening their applications.

REFERENCES

Beatty, R. A. (1974). *Biblphy Reprod.* **23,** 1–6, 127–129.

Bhattacharya, B. C. (1976). *Proc. 8th Int. Cong. Anim. Reprod. Artif. Insem., Krakow, 1976,* **4,** 876–879.

Bhattacharya, B. C., Shome, P., Gunther, A. H., and Evans, B. M. (1977). *Int. J. Fertil.* **22,** 30–35.

Bhattacharya, B. C., Evans, B. M., and Shome, P. (1979). *Int. J. Fertil.* **24,** 256–259.

Bromhall, J. D. (1975). *Nature (London)* **258,** 719–721.

Burfening, P. J., Kress, D. D., Friedrich, R. L., and Vaniman, D. D. (1978). *J. Anim. Sci.* **47,** 595–600.

Eaglesome, M. D., and Mitchell, D. (1977). *Theriogenology* **7,** 195–201.

Ericsson, R. J., Langevin, C. N., and Nishino, M. (1973). *Nature (London)* **246,** 421–424.

Gardner, R. L., and Edwards, R. G. (1968). *Nature (London)* **218,** 346–348.

Hare, W. C. D., and Betteridge, K. J. (1978). *Theriogenology* **9,** 27–43.

Hare, W. C. D., Singh, E. L., Betteridge, K. J., Eaglesome, M. D., Randall, G. C. B., and Mitchell, D. (1978). *Cur. Top. Vet. Med.* **1,** 441–449; 470–472.

Hoppe, P. C., and Illmensee, K. (1977). *Proc. Natl. Acad. Sci. U.S.A.* **74,** 5657–5661.

Markert, C. L., and Petters, R. M. (1978). *Genetics.* **88,** S62–S63.

Modlinski, J. A. (1978). *Nature (London)* **273,** 466–467.

Moustafa, L. A., Hahn, J., and Roselius, R. (1978). *Berl. Muench. Tieraerztl. Wochenschr.* **91,** 236–238.

Pearson, P. L. (1972). *J. Med. Genet.* **9,** 264–275.

Pearson, P. L., Bobrow, M., Vosa, C. G., and Barlow, P. W. (1971). *Nature (London)* **231,** 326–329.

Schneider, U., and Hahn, J. (1979). *Theriogenology* **11,** 63–80.

Short, R. V. (1979). *Brit. Med. Bull.* **35,** 121–127.

Singh, E. L., and Hare, W. C. D. (1977). *Theriogenology* **7,** 203–214.

Singh, E. L., and Hare, W. C. D. (1980). *Theriogenology* **6,** 421–427.

Smith, F. (1919). "The Early History of Veterinary Literature and Its British Development," Vol. 1. Baillière, Tindall & Cox, London.

Soupart, P., Anderson, M. L., and Repp, J. E. (1978). *Theriogenology* **9,** 102.

Sumner, A. T., and Robinson, J. A. (1976). *J. Reprod. Fert.* **48,** 9–15.

Wachtel, S. S. (1979). *Ann. Biol. Anim. Biochim. Biophys.* **19,** 1231–1237.

Willadsen, S. M. (1979). *Nature (London)* **277,** 298–300.

Willadsen, S. M. (1980). *J. Reprod. Fertil.* **59,** 357–362.

Willadsen, S. M., Pashen, R. L., and Allen, W. R. (1980). *Proc. Soc. Study Fertil., Oxford, 1980* (Abstr. No. 32).

Willadsen, S. M., Lehn-Jensen, H., Fehilly, C. B., and Newcomb, R. (1981). *Theriogenology* **15,** 23–29.

7

Preservation of Ova and Embryos by Freezing*

S. P. LEIBO

I. INTRODUCTION

The freezing of biological material is finding increasing application in a variety of disciplines. The reasons are usually rather obvious. The preservation of a population of cells by freezing offers convenience; cells may be collected and then frozen for later retrospective analysis. Preservation by freezing offers a form of biological "insurance"; cells in the frozen state cannot undergo the inevitable genetic drift exhibited by cells growing in continuous culture, nor are the cells subject to possible microbial contamination while in the frozen state. Preservation by freezing also offers new opportunities for medical research. It may be possible, for example, to treat certain forms of cancer by immunotherapy if a tumor removed from a patient could be prepared as a vaccine, frozen, and thawed, and then later used for therapy (Peters and Hanna, 1980). However, a

*The work reported in this article was conducted while the author was a member of the staff of the Biology Division, Oak Ridge National Laboratory, Oak Ridge, Tennessee. ORNL is operated by Union Carbide Corporation under Contract W-7405-eng-26 with the U.S. Department of Energy.

127

TABLE I

Standard Procedure for Cell Preservation

1. Mix equal volumes of cell suspension and 10–15% dimethyl sulfoxide or glycerol.
2. Dispense 0.5–1 ml volumes of cell suspension into glass or plastic freezing vials.
3. Place vials into styrofoam box, and place box in −80°C mechanical refrigerator for ∼2–12 hr.
4. Transfer frozen samples into liquid nitrogen refrigerator at −196°C for storage.
5. For use, thaw frozen samples by shaking them in 37°C water bath for ∼2–3 min. until thawed.
6. Dilute cell suspension ∼10- to 20-fold; wash cells by centrifugation and resuspension in fresh medium. Assay for growth or function.

frequent objection by noncryobiologists is that there do not appear to be "standardized procedures for cryopreservation" but rather a ". . . proliferation of individual recipes, tailored to certain cell types by empirical optimization trials" (Bartlett and Kreider, 1978).

To a certain extent, this perception is justified. Often, freezing procedures for different cell types have been devised solely by empiricism. Alternatively, a single "standardized" procedure has been used to freeze a variety of cell types, yielding excellent preservation of some types but poor to useless survival of other types. Partly for these reasons, cryobiology is sometimes viewed as little more than a technology. Furthermore, from a practical point of view, it seems unnecessary to be concerned with fundamental aspects of cryobiology, since often cells can be successfully preserved by very simple procedures. One example of such a procedure is summarized in Table I. Many types of cells can be successfully frozen in this way, especially if the criterion of "success" is simply the ability of the frozen–thawed cells to exclude vital dyes, e.g., trypan blue or Evans' blue, or to initiate a new cell culture. As an extreme example, consider the case of a population of cells having a mean generation time of ∼ 12 hr, and in which only 0.01% of 10^6 frozen–thawed cells actually survive the preservation procedure. If those 100 survivors are placed into culture, assuming no lag phase of growth, then in ∼ 6½ days the population will again have grown to 10^6 cells. If the mean generation time were ∼ 8 hr and as many as 0.1% of the frozen–thawed cells were to survive, then only ∼ 3⅓ days would be required for the population to grow to 10^6 cells.

II. THE VARIABLES OF CRYOBIOLOGY

If the criteria of successful preservation are more rigorous, however, then the standard procedure shown in Table I may not be sufficient, or too few cells frozen by that procedure may survive. Any one or a combination of the principal variables of cryobiology listed in Table II may be responsible for the lack of successful preservation. First, there is the question of the type and concentration of the protective compound in which the cells are suspended for freezing. A wide variety of such compounds has been demonstrated to protect cells against freezing damage. These compounds are usually divided into two categories, those which freely permeate most cells and those which do not permeate. The former include glycerol, dimethyl sulfoxide, and ethylene glycol; the latter include sucrose, polyvinylpyrrolidone (PVP), hydroxyethyl starch (HES), dextrans, and albumin. Despite a considerable body of literature describing the responses of cells frozen in solutions of these cryoprotective compounds, their fundamental mechanism of action remains unknown.

The second principal variable of cryobiology is ice formation. As a manifestation of the colligative properties of solutions (vapor pressure, osmotic pressure, boiling point, and freezing point), the addition of a solute to water lowers its freezing point below 0°C. The freezing point of a solution depends on the number of molecules of solute present. For example, the freezing point of an aqueous solution of 0.165 M (isotonic) NaCl is about −0.6°C. The freezing points of isotonic NaCl solutions containing 1 M, 2 M, or 3 M glycerol are −1.9°, −4.6°, and −8.1°C, respectively. This means that ice will not form in those solutions until they are cooled below their respective freezing points. However, depending to some extent on the volume being cooled, aqueous solutions also exhibit metastable behavior in which they supercool below their true freezing points with no ice formation. Volumes of ∼ 0.5 ml or less of aqueous solutions can often be supercooled some 10 to 15 degrees below their freezing points. This phenomenon of supercooling may play a critical role in the ultimate survival of cells being frozen, especially mammalian ova and embryos. The reason follows from a fundamental characteristic of all types of mammalian cells. This characteristic is the osmotic behavior of cells. When a cell is transferred from an isotonic solution into a hypertonic one, it responds osmotically by losing water to achieve equilibrium between its intracellular solution and the extracellular one in which it is suspended. The rate at which a given cell type equilibrates osmotically is defined

TABLE II

Principle Variables of Cryobiology

1.	Type and concentration of cryoprotective compounds
2.	Ice formation
3.	Cooling rate from ∼ −5° to −70°C
4.	Storage temperature
5.	Warming rate from ∼ −70° to −5°C
6.	Dilution rate and temperature

by its hydraulic conductivity (L_p) or its water permeability coefficient. This L_p reflects the cell's surface area to volume ratio and is principally influenced by temperature and the difference in osmotic pressure between the intracellular and extracellular solutions. Different types of cells may possess drastically different hydraulic conductivities. For example, most mammalian erythrocytes have L_p's of ~ 5 to 8 $\mu m^3 / \mu m^2$-min-atm at 20°C (see tabulation in House, 1974), whereas many leukocytes have L_p's of ~ 0.3 to 0.4 (DuPré and Hempling, 1978). Mouse ova also have an L_p of ~ 0.4 $\mu m^3/\mu m^2$-min-atm at 20°C (Leibo, 1980). Of equal importance is the effect of temperature on L_p, that is, the temperature coefficient of water permeability. Regardless of the units in which this temperature coefficient is expressed, a reduction in temperature causes a corresponding reduction in the water permeability of a cell. Different types of cells exhibit different temperature coefficients of water permeability. For example, erythrocytes have low temperature coefficients (House, 1974); mammalian ova have high coefficients (Leibo, 1980). This means that, in relative terms, a rather small reduction in temperature drastically reduces the rate at which an ovum can lose water, whereas a rather large reduction in temperature causes only a small decrease in the rate at which an erythrocyte can lose water.

The interaction of the supercooling of a solution and the water permeability of a given cell type may determine whether or not that cell survives cooling to lower subzero temperatures. Consider the diagrams in Fig. 1. Panels A and B show the sample and bath temperatures of two identical samples being cooled at the same rate. Panels C and D show diagrammatically the corresponding response of a cell in each sample. The sample on the left supercools, i.e., no ice forms. The sample on the right freezes at about −2°C, manifested by an increase in the sample

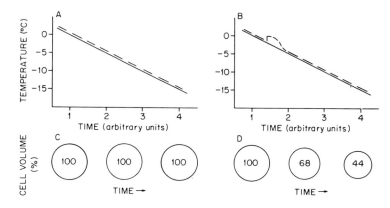

Fig. 1. Diagram of the temperature record of samples cooled at the same rate either without having been seeded (A) or after having been seeded (B) to induce ice formation. The lower panels show diagrammatically the changes produced in the volume of a cell in the corresponding samples (C, D).

temperature caused by the release of the latent heat of fusion as the ice formed. The formation of this ice effectively removes water from the solution, causing a concomitant increase in the solution concentration. The cell, suddenly exposed to a more concentrated solution, responds osmotically by losing water, i.e., decreasing its volume. As that solution is cooled to a lower temperature, more ice forms causing a further increase in the solution concentration with a corresponding further decrease in the cell volume. There is one critically important proviso to that description of the cell's response. The cooling rate must be low enough so that, as a function of the cell's permeability to water and the temperature coefficient of water permeability, there is sufficient time for the cell to respond osmotically. The qualitative expression "sufficient" can be quantified by a series of mathematical equations derived by Mazur (1963). More importantly, Mazur's equations can also be used to describe the entire response of a cell during freezing. This aspect will be discussed in more detail below. Turning back to the hypothetical cell shown in panel C, it should now be clear that, since no ice formed in the first sample (panel A), there would be essentially no change in the concentration of the solution. Therefore, there would be no driving force to cause the cell in a supercooled solution to change its volume by losing water. Thus, at some subzero temperature, for instance, $-15°C$, a cell in a supercooled solution would contain the same volume of water that it did at $0°C$, whereas a cell in a partially frozen solution at the same temperature would have lost about one-half of its initial volume. These two "states" of the cell (fully hydrated versus half dehydrated) will largely determine the ultimate fate of a cell when it is cooled and frozen to low subzero temperatures.

The third principal cryobiological variable, cooling rate, then begins to exert its effects. It has long been recognized that a given cell can be cooled at rates that are too low or too high for it to survive freezing and thawing. Fortunately, there also exist intermediate or "optimum" cooling rates that do yield high survival of many types of cells. Mazur (1963, 1966) was the first to recognize that different types of cells exhibit different optimum cooling rates. More importantly, he offered a quantitative explanation of those differences. His equations referred to above provide the explanation (see Mazur [1963], pp. 350-353).

The data shown in Fig. 2 illustrate the problem. When mouse ova, mouse lymphocytes, or human erythrocytes are suspended in solutions of dimethyl sulfoxide (Me_2SO) and then are cooled to low subzero temperatures, their survival depends critically on the cooling rate. About 70% of the ova survive when cooled at rates of ~ 0.4-$1.5°C/min$; none survives when cooled at $5°C/min$ or faster. Almost 60% of the lymphocytes survive when cooled at $\sim 15°C/min$; few if any survive when cooled at rates about 15 times faster or slower. More than 90% of the erythrocytes survive when cooled at $\sim 1000°C/min$; fewer than 20% survive when cooled at rates 6 times faster or about 20 times slower. From a practical point of view, these results show that three disparate types of cells can

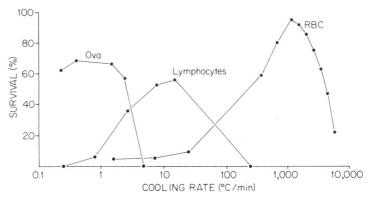

Fig. 2. Comparison of the survival of three types of cells suspended in dimethyl sulfoxide as a function of cooling rate. The data for ova are those of Leibo *et al.* (1978), for lymphocytes of Thorpe *et al.* (1976), and for RBC are a composite constructed from data of Morris and Farrant (1972) and of Rapatz and Luyet (1965).

indeed be preserved by freezing if cooled at an optimum rate. But from a fundamental viewpoint, how can one explain the 1000-fold difference in optimum rates for ova and erythrocytes?

The answer is found principally in the difference between the water permeability coefficients of different cells and their temperature coefficients of water permeability described above. Consider the data shown in Table III for three types of cells. Ova and lymphocytes have low water permeabilities and large activation energies; these characteristics tend to reduce the rate at which these cells lose water with decreasing temperature. However, ova have a diameter of $\sim$ 75 μm and a volume of $\sim$ 2.2 $\times$ 10^5 μm^3; lymphocytes have a diameter of $\sim$ 15 μm or less and a volume of $\sim$ 1.8 $\times$ 10^3 μm^3. Therefore, the ratio of surface area to volume (SA/V) of lymphocytes is some 7.5 times larger than that of ova. This means that lymphocytes have a much larger relative surface across which water

TABLE III

Fundamental Characteristics of Three Mammalian Cell Types

Cell type	L_p (μm^3/μm^2-min-atm)	ΔH^+ (Kcal/mole)	SA/V	Optimum cool. rate (°C/min)
Mouse ova	0.4	$\sim$ 14	0.08	$\sim$ 1
Mouse lymphocyte	0.4	13–18	0.6	15
Human erythrocyte	5.7	3–5	1.9	1200

can move than do ova. In contrast to these characteristics, erythrocytes have a high water permeability, a small activation energy, and, because of their biconcave discoid shape, a large SA/V. Taken together, these characteristics of erythrocytes result in rapid water movement even with decreasing temperature. These fundamental characteristics (L_p, ΔH^+, and SA/V) are principal determinants of the cells' responses to cooling rate. The optimum cooling rate for ova is $\sim 1°C/min$, for lymphocytes $\sim 15°C/min$, and for erythrocytes $\sim 1200°C/min$. Mazur's (1963) mathematical analysis of the kinetics of water loss at subzero temperatures predicted that cellular characteristics that tended to reduce the rate of water loss with decreasing temperature would also reduce the cooling rate at which ice would form inside a cell. Comparison of the fundamental characteristics of the three cell types shown in Table III with their corresponding optimum cooling rates confirms Mazur's prediction. A cell type that loses water slowly in response to an osmotic pressure gradient, e.g., an ovum, has a low optimum cooling rate. A cell type that loses water rapidly, e.g., an erythrocyte, has a high optimum cooling rate. And a cell type that loses water at an intermediate rate, because of its relatively larger SA/V, e.g., a lymphocyte, has an intermediate optimum cooling rate.

The fourth variable of cryobiology is the temperature at which the frozen cells are stored. In general, two storage temperatures have been used, $-80°$ and $-196°C$. The former is approximately the sublimation temperature of Dry Ice and is that obtained in many mechanical laboratory freezers. In some special situations, e.g., the storage of human erythrocytes frozen in high concentrations of glycerol (Huggins, 1963), the frozen cells are stable for periods of about 1 year. In general, however, cells to be preserved for extended times are stored below $-130°C$, the "glass transition" temperature of ice. The most reliable, least expensive, and safest way to accomplish this is to store the frozen cells either in the liquid ($-196°C$) or vapor ($-150°C$) phase of liquid nitrogen. Despite occasional reports suggesting loss of biological activity at $-196°C$, none has been unequivocally documented. There are, on the other hand, numerous studies demonstrating full retention of viability of a wide variety of cells stored for years at $-196°C$. For example, Lyon *et al.* (1981) have recently shown that mouse embryos frozen for > 5 years exhibit the same *in vitro* and *in vivo* survival as embryos frozen for only 24 hr. Sherman (1973) has reported the birth of children resulting from artificial insemination with sperm stored for more than 10 years. As long as care is exercised to insure that the cells are maintained below $-150°C$ with no interruption, frozen cells ought to be stable for decades or longer.

Warming rate, the fifth principal variable, may exert as profound an effect on cell survival as does cooling rate. In general, there is a strong interaction between the effects of cooling rate and warming rate on cell survival. One example for 8-cell mouse embryos is shown in Fig. 3. When embryos suspended in a Me_2SO

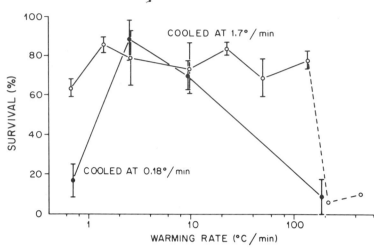

Fig. 3. Survival of mouse embryos as a function of the rate at which they were warmed after having been frozen at either 0.18° or 1.7°C/min to −196°C. The figure itself is from Leibo *et al.* (1974) and is reproduced with permission of the publisher.

solution are cooled at a low rate of ∼ 0.2°C/min, then their survival is strongly dependent on the rate at which they are warmed. Ninety percent survive when the frozen embryos are warmed at ∼ 3°C/min; less then 20% survive when they are warmed very slowly or very rapidly. When embryos are cooled about 10 times faster at 1.7°C/min, then their survival is rather independent of warming rate, unless they are warmed very rapidly at 200°C/min or faster. These results indicate that, while still frozen at −196°C, the state of an embryo that had been cooled at ∼ 0.2°C/min is drastically different from one that had been cooled at ∼ 2°C/min. Qualitatively, these results may also be explicable in terms of water permeability, just as were the effects of cooling rate. An embryo cooled slowly has sufficient time to dehydrate during cooling. However, an embryo cooled more rapidly has insufficient time to dehydrate. When a frozen embryo is warmed, it is subjected to a dilution of its suspending media as the ice begins to melt. That is, a solution of Me_2SO will begin to melt at ∼ −70°C. At that temperature, the concentration of the slightly thawed solution will be at least 10 *M*. If the original concentration of the Me_2SO was 1 *M,* then the solution will obviously return to that concentration when it is completely melted. Inevitably, then, an embryo frozen in 1 *M* Me_2SO will be exposed to very high concentrations of solute at low temperatures and will be exposed to decreasing concentrations as the suspension is warmed to higher and higher temperatures. When a frozen embryo is warmed and the extracellular solution begins to melt, the osmotic properties of the embryo will require it to attempt to equilibrate with the melting solution. If the warming rate is rather moderate, then the embryo apparently can remain in approximate osmotic equilibrium as it is warmed. If the warming rate is high, however, then the embryo must be required to undergo a large and rapid rehydration during warming. The extent of this rehydration will depend on the extent to which the embryo has dehydrated during the cooling procedure. The slower the cooling, the more the embryo will have dehydrated, and the more it will have to rehydrate during warming. Apparently, if the rehydration is too large and too rapid, then the embryo is killed. It may be that such large rapid volume changes are damaging to embryos if they occur at subzero temperatures. This still leaves unexplained, however, why very slow cooling, for instance at $0.2°C/min$, followed by very slow warming, for instance at $0.7°C/min$, is also a damaging sequence. There is no satisfactory explanation of that phenomenon at this time.

The final principal variable of cryobiology is dilution rate and temperature. This refers to the rate and temperature at which the cell or embryo is diluted out of the solution in which it was originally frozen. For mammalian embryos to survive freezing, they must be suspended in a solution of a cryoprotective solute. Those that have been demonstrated to protect embryos against freezing damage include dimethyl sulfoxide or glycerol (Whittingham et al., 1972), and various glycols (Miyamoto and Ishibashi, 1978; and Renard et al., 1981). Among these various compounds, only glycerol has been explicitly demonstrated to permeate mammalian ova and embryos (Jackowski et al., 1980). It can be reasonably inferred, however, that dimethyl sulfoxide and low-molecular-weight glycols do, in fact, permeate mammalian embryos. For frozen–thawed embryos to develop either in vitro or in vivo, they must be washed free of the cryoprotective compound. This is a step that is often overlooked in the freezing of single-cell suspensions consisting of 10^6 cells or more, since some of the cells will survive even when diluted directly out of the protective solution. With embryo freezing, however, care must be paid to dilute the protective compound in such a way as to avoid an osmotic shock. This shock results if the embryo, containing a rather high concentration of protective solute, is suddenly transferred into an isotonic salt solution. Water will move into the embryo more rapidly than the solute can flow out. The embryo will swell to attempt to achieve osmotic equilibrium. Even though the embryo is usually surrounded by the zona pellucida, the swelling is often sufficient to rupture the cells of the embryo itself. Numerous methods have been devised to avoid this osmotic shock of embryos (Leibo and Mazur, 1978). One is to dilute the extra embryonic solution by a series of slow step-wise dilutions. Another is to raise the temperature at which the embryo is diluted. This increases the rate at which the protective solute flows out of the embryo, thus reducing the difference in osmotic pressures between the intracellular and extracellular solutions.

In summary, each of the principal variables of cryobiology affects the viability of mammalian ova and embryos. For the embryo to survive freezing and thawing, it must survive each of the potentially lethal consequences of all six of these variables. It should also be clear that the successful preservation of ova and embryos requires that the embryo survive the entire concatenation of "insults" from (1) exposure to an extremely nonphysiological solution, (2) phase transformation of the suspending solution, (3) cooling to, (4) storage at, and (5) warming from extremely low subzero temperatures, and finally, (6) return to physiological conditions.

III. PRACTICAL ASPECTS OF EMBRYO PRESERVATION

From the beginning, experiments to preserve mammalian embryos by freezing have had a dual purpose. First, there are the practical reasons. Embryo preservation provides a means by which mammalian species can be stored for extended times with the real potential to re-establish breeding populations at a later time. This is particularly valuable when applied to the preservation of mutant strains of mice or other laboratory species. When applied to large domestic animals, such as cattle, sheep, or horses, embryo preservation drastically increases the efficiency of embryo transfer to improve the genetic selection of valuable individual animals. The second purpose of embryo freezing has been to study fundamental aspects of cryobiology. For example, in the first published report of embryo freezing that yielded live offspring, Whittingham *et al.* (1972) concluded that ". . . The success of cryobiological theory in suggesting the proper approach to the freezing of these sensitive embryos increases the likelihood that ways can be found to freeze complex mammalian systems for medical use." This has, indeed, proved to be true. In 1976, Mazur et al. applied the principles discovered from embryo freezing to pieces of whole animal tissue. They showed that similar freezing procedures could be used to preserve the functional survival of fetal rat pancreases. Other experiments on embryo freezing have been applied to even more basic aspects of cryobiology (e.g., Leibo *et al.*, 1974; Leibo *et al.*, 1978; Rall *et al.*, 1980). The reason for this latter situation is partly that mammalian ova and embryos are so large. (The bovine blastocyst has a diameter of $\sim$ 150 μm; the bovine erythrocyte has a diameter of $\sim$ 6 μm.) Many microscopic observations are simply easier to perform on larger cells or collections of cells. Nevertheless, the primary reason for interest in embryo freezing stems from its practical applications.

Because of the practical uses of embryo preservation, much effort has been devoted to simplifying the methods used to freeze embryos. Table IV summarizes the present procedures. All of these are explicable in terms of the funda-

TABLE IV

Comparison of Methods to Freeze Embryos

Method	Cooling			Warming		Total time (min)	References
	Rate (°C/min)	Temperature range (°C)	Time required (min)	Rate (°C/min)	Time required (min)		
I	~0.5	−5° to <−70°	130	~20	10 to 15	145	Whittingham et al. (1972)
	>300	−80° to −196°	<0.5				
II	0.3	−6° to −30°	80	~360	~0.5	111	Willadsen et al. (1977)
	0.1	−30° to −33°	30				
	>300	−33° to −196°	<0.5				
III	~7	−5° to −20°	~2	~360	~0.5	28	Kasai et al. (1980)
	—	Hold at −20°	10				
	17	−20° to −100°	5				
	—	Hold at −100°	10				
	>300	−100° to −196°	<0.5				
IV	0.3	−9° to −40°	103	~500	~0.4	104	Whittingham et al. (1979)
	>300	−40° to −196°	<0.5				

mental cryobiology described in Section II. Method I was the original one derived in the first experiments on this subject. Briefly, it involves cooling the embryos slowly to a rather low subzero temperature, storing them in liquid nitrogen at $-196°C$, and then warming them slowly until thawed. The total time required for freezing and thawing is about 2½ hr. Method II, developed by Willadsen et al. (1977), requires that the embryos only be cooled slowly to the relatively high temperature of $\sim -33°C$, thus reducing the time required for cooling. It also requires that the frozen embryos be warmed rapidly. This method was primarily developed to freeze embryos of large animals, such as cattle and sheep. It is often referred to as the "short program" for embryo freezing. However, it can be easily seen from the figures in Table IV that it actually reduces the total time for freezing and thawing by only about 30 min compared to Method I. Rather recently, a variation on the theme of "short" freezing was introduced by Whittingham et al. (1979), Method IV. The difference in time required between it and Method I approaches 40 min. The most significant improvement in embryo freezing procedures, from the point of view of time required, is that of Kasai et al. (1980), Method III. This method only requires about 30 min for the entire freeze–thaw process. Wood and Farrant (1980) have reported similar findings for the so-called "two-step" freezing of mouse embryos. Since many other types of cells have been demonstrated to survive two-step freezing, this is yet one more presumptive example that fundamentals of cryobiology apply with equal validity to mammalian embryos and a variety of other types of cells.

In addition to the actual methods used to freeze and thaw embryos, various other modifications have been employed for embryo freezing. For example, Zeilmaker and Verhamme (1979) have shown that mouse embryos can be frozen while still contained within the oviduct itself. This reduces the time for embryo freezing by eliminating the need to collect the embryos before freezing them. That means that more embryos can be frozen and stored. This method has particular value for mouse embryo banking, since it can be anticipated that more mouse strains will be frozen than will be thawed. Only those strains to be re-established as breeding colonies will need to be thawed and transferred. Of course, embryos frozen in this way still must be collected from the thawed oviducts in order to be transferred. Another potentially practical modification of embryo freezing is the use of plastic straws as containers for the embryos. These straws are routinely used to freeze bovine semen for artificial insemination. Tsunoda and Sugie (1977) and Massip et al. (1979) have shown that embryos can be successfully frozen in such straws. At the present time, however, the embryos must be recovered from the straws in order to be diluted out of the dimethyl sulfoxide in which they were frozen. Ultimately, it would be most advantageous if bovine embryos could be frozen, thawed, and diluted within the straw itself. This would mean that bovine embryo transfer could be performed as simply and quickly as bovine artificial insemination, an obvious advantage for efficient husbandry of cattle.

IV. CONCLUSION

The preservation of mammalian ova and embryos by freezing has progressed rapidly since its first successful demonstration in 1972. Both from the points of view of fundamental cryobiology and of embryo transfer, experiments on embryo freezing have contributed to a new technology in animal breeding. It can be reasonably anticipated that experiments in this area will continue to yield fundamental understanding of biological processes and more practical means to increase domestic animal productivity.

REFERENCES

Bartlett, G. L., and Kreider, J. W. (1978). *Cancer Immunol. Immunother.* **3,** 213.

DuPré, A. M., and Hempling, H. G. (1978). *J. Cell. Physiol.* **97,** 381–395.

House, C. R. (1974). "Water Transport in Cells and Tissues." Williams and Wilkins, Baltimore, Maryland.

Huggins, C. E. (1963). *Surgery* **54,** 191.

Jackowski, S. C., Leibo, S. P., and Mazur, P. (1980). *J. Exp. Zool.* **212,** 329–341.

Kasai, M., Niwa, K., and Iritani, A. (1980). *J. Reprod. Fertil.* **59,** 51–56.

Leibo, S. P. (1980). *J. Membr. Biol.* **53,** 179–188.

Leibo, S. P., and Mazur, P. (1978). *In* "Methods in Mammalian Reproduction" (J. C. Daniel, Jr., ed.), pp. 179–201. Academic Press, New York.

Leibo, S. P., Mazur, P., and Jackowski, S. C. (1974). *Exp. Cell Res.* **89,** 79–88.

Leibo, S. P., McGrath, J. J., and Cravalho, E. G. (1978). *Cryobiology* **15,** 257–271.

Lyon, M., Glenister, P., and Whittingham, D. G. (1981). *In* "Frozen Storage of Laboratory Animal Embryos" (G. H. Zeilmaker, ed.). Gustav Fischer Verlag, Stuttgart (in press).

Massip, A., van der Zwalmen, P., Ectors, F., De Coster, R., D'Ieteren, G., and Hazen, C. (1979). *Theriogenology* **12,** 79–83.

Mazur, P. (1963). *J. Gen. Physiol.* **47,** 347–369.

Mazur, P. (1966). *In* "Cryobiology" (H. T. Meryman, ed.), pp. 2J3–315. Academic Press, New York.

Mazur, P., Kemp, J. A., and Miller, R. H. (1976). *Proc. Natl. Acad. Sci.* **73,** 4105–4109.

Miyamoto, H., and Ishibashi, T. (1978). *J. Reprod. Fertil.* **54,** 427–432.

Morris, G. J., and Farrant, J. (1972). *Cryobiology* **9,** 173–181.

Peters, L. C., and Hanna, M. G., Jr. (1980). *J. Natl. Cancer Inst.* **64,** 1521–1525.

Rall, W. F., Reid, D. S., and Farrant, J. (1980). *Nature (London)* **286,** 511–514.

Rapatz, G., and Luyet, B. (1965). *Biodynamica* **9,** 333–350.

Renard, J.-P., Heyman, Y., and Ozil, J.-P. (1981). *Theriogenology* **15,** 113.

Sherman, J. K. (1973). *Fert. Steril.* **24,** 397–412.

Thorpe, P. E., Knight, S. C., and Farrant, J. (1976). *Cryobiology* **13,** 126–133.

Tsunoda, Y., and Sugie, T. (1977). *J. Reprod. Fertil.* **49,** 173–174.

Whittingham, D. G., Leibo, S. P., and Mazur, P. (1972). *Science (Washington, D.C.)* **178,** 411–414.

Whittingham, D. G., Wood, M., Farrant, J., Lee, H., and Halsey, J. A. (1979) *J. Reprod. Fertil.* **56,** 11–21.

Willadsen, S. M., Polge, C., and Rowson, L. E. A. (1977). *J. Reprod. Fertil.* **52,** 391–393.

Wood, M., and Farrant, J. (1980). *Cryobiology* **17,** 178–180.

Zeilmaker, G. H., and Verhamme, C. M. P. M. (1979). *Cryobiology* **16,** 6–10.

8

Applications of *in Vitro* Fertilization

BENJAMIN G. BRACKETT

I. INTRODUCTION

The technology of *in vitro* fertilization and embryo culture is contributing to improved approaches to inhibit and enhance reproductive efficiency of mammals. Progress in understanding sperm–egg union has followed development of *in vitro* fertilization experimentation in several laboratory species, especially the rabbit, mouse, rat, hamster, and guinea pig. Evidence of fertilization *in vitro* has

141

NEW TECHNOLOGIES IN ANIMAL BREEDING
Copyright © 1981 by Academic Press, Inc.
All rights of reproduction in any form reserved.
ISBN 0-12-123450-9

also been reported for nine other mammalian species including man, cat, dog, pig, sheep, and cow. However, when coupled with embryo transfer, normal gestational development has only been reported in rabbit, mouse, rat, cow, and human experiments. Documentation of repeatable procedures yielding high percentages of fertilizations and viable embryos is limited to laboratory animal species. An inadequate understanding of the development of fertilizing ability by sperm cells, e.g., capacitation and the acrosome reaction remains a major technical obstacle to repeatability. *In vitro* culture of the zygote and embryo represents an essential link between *in vitro* fertilization and embryo transfer. Improvements are necessary to allow embryonic development to proceed *in vitro* as it does normally in the oviduct. At this writing, it has been possible to circumvent the oviduct with *in vitro* fertilization and embryo transfer only in mouse and man. Rapid progress in research is anticipated and findings should become increasingly applicable to animal breeding within the next 10–20 years.

II. RESEARCH

The applicability of *in vitro* fertilization methodology, especially during the last decade, has greatly facilitated progress in our understanding of gamete physiology, fertilization, and *in vitro* culture. (For reviews, see Austin, 1974; McRorie and Williams, 1974; Whittingham, 1975, 1979; Chang *et al.*, 1977; Gwatkin, 1977; Yanagimachi, 1977, 1981; Meizel, 1978; Rogers, 1978; Brackett, 1979b, 1980, 1981; Bavister, 1980, 1981). Research involving gametes of laboratory animal species has resulted in exquisite morphological descriptions of fertilization. Conditions that enable gamete union *in vitro* have been adequately established to facilitate the unraveling of many biochemical events.

A. Conditions Necessary for the Ovum

Emphasis has been placed on defining conditions that allow fertilization to occur *in vitro*. Optimal results are achieved when the ovum (also, egg or oocyte) to be fertilized is recovered just prior to, at the time of, or just after ovulation has taken place. At this time, the egg is in the second metaphase of meiosis and hence its maturation is complete. An abundance of follicular cells surrounding the ovum is a characteristic feature at this stage. The correct timing of ovum recovery in primates must be guided by endocrine profiles. Especially useful are rapid assays to detect the ovulation-dependent surge of luteinizing hormone (LH). In most lower animals, ovulation can be timed within acceptable limits by a combination of behavioral events and hormonal treatments (for superovulation). In the mare and cow, follicular development can be assessed by palpation per rectum. In the mare, superovulation is not yet practical. For many applications, streamlined methodology for oocyte recovery, e.g., aspiration of mature follicle(s) via laparoscopy (Steptoe and Edwards, 1970; Dukelow, 1978), will become necessary. The unfertilized ovum is extremely vulnerable to fluctuations in temperature and other environmental conditions. It is, therefore, important to provide a milieu compatible with maintenance of the functional life of the female gamete prior to union with the sperm cell. This has been accomplished by conditions simulating the environment of the oviduct with fluids containing physiological concentrations of salts, hydrogen ions (pH of 7.6–7.8), protein, and substrates for metabolic maintenance. Maintenance of body temperature and a reduced oxygen tension (5–8%) are of special importance. Conditions that best support fertilization have to be altered somewhat for optimal culture of the zygote and embryo, again consonant with the dynamic milieu of the oviduct (see Chapter 3, this volume).

B. Preparation of Spermatozoa

In contrast to the ovum, which is capable of undergoing fertilization at the time of ovulation, the sperm cell must undergo the process of capacitation following ejaculation. This normally occurs in the female reproductive tract and has been duplicated in a crude way *in vitro* by washing procedures or treatment with a hypertonic solution and/or preincubation for varying intervals before exposure to ova, depending upon the species under study. After capacitation, a change in sperm membranes, the acrosome reaction, allows release of enzymes important in penetrating the egg. Experimental treatments to effect sperm capacitation and/or the acrosome reaction have frequently included *in vivo* incubations or incubation with undefined biological substances, e.g., follicular fluid and serum. Conditions to enable sperm cells to penetrate ova *in vitro* have been found, with few exceptions through experimental trial and error. Recent progress has been made toward lessening the formidable task presented by such trial and error. Now, the successive removal or alteration of sperm-coating antigenic components of seminal plasma that reflects the temporal development of sperm fertilizing ability can be monitored (Oliphant and Brackett, 1973; Brackett and Oliphant, 1975).

A promising approach to assessment of capacitation and the acrosome reaction developed for human sperm involves *in vitro* insemination of hamster ova that have been enzymatically divested of the follicular cells and zonae pellucidae (Yanagimachi *et al.*, 1976). Only those sperm completing these prerequisites can interact with zona-free hamster ova to form male pronuclei. Penetration of zona-free hamster ova *in vitro* has also been reported for sperm of the boar (Imai *et al.*, 1977, 1979) and of the bull (Lorton and First, 1979; Brackett *et al.*, 1980a; Bousquet and Brackett, 1981). In the writer's laboratory this test has been useful as a guide in defining conditions that enable bull sperm to undergo capacitation and the acrosome reaction. The incomplete understanding of appropriate sperm conditioning, which is complicated by biological variability among individual males, represents a major reason for the current inability to consistently accomplish fertilization *in vitro* for most species.

C. Viability of *in Vitro* Fertilized Ova

Even though a sperm has penetrated the zona pellucida (which represents the major barrier presented by the ovum), there is no assurance that a resulting zygote has the potential for continued development into a new individual. A major concern associated with *in vitro* exposure of ova to spermatozoa is that *in vivo* screening or selection of the most suitable sperm cell(s) for fertilization is not operative. Under such circumstances, a greater proportion of zygotes and early embryos that are weak, incompetent, or abnormal in some way might be expected. However, it is estimated that around 20–40% embryonic mortality normally occurs in domestic animal species, and even higher losses in man. Most of the embryonic wastage probably arises from chromosomal aberrations from the sperm and/or ovum (for reviews, see Short, 1979b; Biggers, 1981). Cytogenetic studies of products of *in vitro* fertilization will show whether such losses are magnified following *in vitro* fertilization. A higher than normal incidence of polyspermy occurs in hamster *in vitro* fertilization experiments (Barros and Yanagimachi, 1972), and this undoubtedly contributes to the uniformly unsuccessful efforts to obtain gestational development after transfer of *in vitro* fertilized hamster ova. Fertilization by an X-bearing sperm of a human ovum with a nonfunctional nucleus followed by regulation to diploidy may lead to development of a hydatidiform mole, which in turn might lead to choriocarcinoma (Kajii and Okama, 1977). Caution has been raised regarding the possibility of choriocarcinoma as a result of human *in vitro* fertilization efforts, but, on the positive side, an opportunity might be presented for further study of initial events in the pathogenesis of this form of cancer if these early abnormal events could be reproduced *in vitro* (Short, 1979a, b; Soupart, 1979a, b).

Genetically controlled enhancement of *in vitro* gamete union occurs in experiments with various strains of mice (Parkening and Chang, 1976a, b). Several observations with mice along these lines during the past 5 years have led to recognition of genetic superiority of sperm, ova, and/or embryos regarding facility to unite and/or develop *in vitro* (for review, see Brackett, 1981). Also, development to the pronuclear stage following sperm penetration of ova *in vitro* is dramatically influenced by state of maturity of the ova (Thibault *et al.*, 1975; Fraser, 1979). In an experimental setting, *in vitro* fertilization can provide an endpoint for basic research on sperm maturation and on oocyte maturation as well.

Observations on the developmental potential following embryo transfer of products of *in vitro* fertilization are not extensive (Table I). After fertilization and initial ovum cleavage, *in vitro* fertilized embryos should be comparable to *in vivo* fertilized embryos in their ability to survive more adverse environmental conditions, e.g., room temperature. The incidence of successful term development of *in vitro* cultured rabbit embryos was similar whether sperm penetration took place *in vitro* or *in vivo* (Mills *et al.,* 1973). The low percentage of *in vitro* fertilized rabbit embryos that developed into live young (around 18%) suggests a need for improvement of culture conditions, but other explanations (e.g., too few developing conceptuses for normal pregnancy maintenance) are also valid. Good progress, especially in the mouse, has been made in studies involving culture of the zygote and embryo (for review, see Brackett, 1981). Although details were not included, Whittingham (1975) reported the highest survival after transfer (a) of 2-cell stage mouse embryos resulting from *in vitro* fertilization to be 54% or 42/78 (Kaufman and Whittingham, cited by Kaufman, 1973) and (b) of blastocysts to be 60% or 49/81. When survival to term was estimated from the original population of oocytes with which fertilization *in vitro* was attempted the highest rates of survival from transfers at the 2-cell and blastocyst stages were similar, 44% versus 49%. Whittingham (1975) suggested that the improved survival of transfers at the blastocyst stage might reflect improved culture conditions where low oxygen (5%) was used for zygote culture (Hoppe and Pitts, 1973; D. G. Whittingham, unpublished observations), but also noted that the best survival rates did not reach those observed after transfer of embryos resulting from *in vivo* development 65% or 85/131 in the best experiments of Mullen and Carter (1973).

Few experimental designs have focused on possibilities of increased abnormal development (Hall and Goulding, 1981); available data support the widely held view that the embryo is usually not susceptible to teratogenesis in the predifferentiation period (Wilson, 1973; Hall, 1981). Clearly, greater embryonic wastage follows *in vitro* fertilization and embryo transfer, but with present technology, many factors probably contribute to this. Higher proportions of offspring should result from transfer of only morphologically normal embryos analogous to efforts in transfer of *in vivo* fertilized bovine embryos (see Chap. 3, this volume). With over 500 pregnancies reported from animal experiments, there is no compelling evidence that the incidence of abnormal offspring might be increased as a result of specific conditions imposed through *in vitro* fertilization procedures. Two of 37 rabbits had splay-legs in one study (Fraser and Dandekar, 1973a) and in rat experiments (11 of 43 offspring) had microphthalmia (Toyoda and Chang, 1974). In these studies, it is most likely that the genetic stock was responsible for the observed defects. In view of recent efforts to apply *in vitro* fertilization to overcome human infertility, emphasis on experimentation in a wide array of animal species would appear to be appropriate in providing better information regarding possible risks of abnormal development.

Successes and failures in human *in vitro* fertilization were recently summarized by Lopata (1980). Three infertile women have given birth to normal infants following *in vitro* fertilization and embryo transfer. In each case, a single

TABLE I

Gestational Development of *in Vitro* Fertilized Ova

Species	Ova	Sperm	Stage reached in culture	References
Rabbit	Oviductal	Uterine	4-cell	Chang (1959), Thibault and Dauzier (1961), and Bedford and Chang (1962)
			8-cell	Brackett (1969)
			2-cell to 4-cell	Fraser et al. (1971) and Fraser and Dandekar (1973a,b)
			4-cell to 8-cell (stored frozen)	Schneider et al. (1974)
			Morulae and early blastocysts	Seidel et al. (1976)
	Ovarian surface	Uterine	4-cell	Seitz et al. (1970) and Mills et al. (1973)
			4-cell (stored 24 h at 10°C)	Jeitles and Brackett (1974)
			2-, 4-, and 8-cell	Brackett (1978a,b)
	Follicular	Uterine	2-, 4-, and 8-cell	Brackett et al. (1972)
			4-cell	Mills et al. (1973)
	Ovarian surface	Ejaculated	2- and 4-cell	Brackett and Oliphant (1975)

Mouse	Oviductal		2-cell	Whittingham (1968)
		Uterine	Blastocyst	Mukherjee and Cohen (1970)
			2- and 4-cell	Kaufman and Whittingham (1972)
	?		2-cell, blastocyst	Kaufman (1973), Whittingham (1975)
		Epididymal	2-cell	Miyamoto and Chang (1972)
			Morula and early blastocyst	Hoppe and Pitts (1973)
	Oviductal (stored frozen)		Blastocyst	Fraser and Drury (1975)
	Follicular		Blastocyst	Parkening et al. (1976)
			2-cell, morula and early blastocyst	Whittingham (1977)
		Uterine	2-cell	Cross and Brinster (1970)
		Epididymal	Blastocyst	Mukherjee (1972)
			Blastocyst	Hall and Goulding (1981)
Rat	Oviductal	Epididymal	2-cell	Toyoda and Chang (1974)
Man	Follicular	Ejaculated	8-cell	Steptoe and Edwards (1978) and Edwards et al. (1980), and Steptoe et al. (1980)
			8-cell	Lopata et al. (1980)
Cow	Follicular	Ejaculated	4-cell	Brackett et al. (1981)

TABLE II

In Vitro **Fertilization in Large Domestic Animals**

| Species | Description of best conditions used | | | Criteria for achievement of fertilization | References |
	Ova	Sperm	Media		
Cow	Follicular oocytes	Frozen sperm, incubated *in utero*	Ringer solution or TC medium 199 + fetal calf serum	2 or more polar bodies, cleavage stages (2- to 9-cell)	Bregulla *et al.* (1974)
	Follicular matured *in vitro*	Ejaculated sperm incubated in excised (estrous) oviduct and uterus	Modified KRB	Sperm penetration and pronuclear development	Iritani and Niwa (1977)
	Follicular and tubal	Ejaculated and high ionic strength (HIS) treated	Defined medium, then modified Ham's F-10	Loss of cortical granules and cleavage to 4-cell stage	Brackett *et al.* (1977, 1978b, 1980c)

Species	Oocyte source	Sperm source	Medium	Result	Reference
	Follicular	Same + preincubation	Defined medium	Sperm within perivitelline space or ooplasm	Brackett et al. (1980b)
	Follicular and tubal	Same	Same, then 10% serum solution, transfer into recipient	Sperm penetration and pronuclear development, cleavage to 4-cell stage and gestational develop.	Brackett et al. (1981)
Pig	Follicular, matured in vitro	Ejaculated sperm preincubated in excised (estrous) uterus	Modified KRB	Cleavage to 4-cell stage	Iritani et al. (1974)
Sheep	Oviductal	Uterine sperm	Locke solution	2 polar bodies and pronuclear development	Dauzier and Thibault (1959)
	Same	Same	Modified KRB	Same	Thibault and Dauzier (1961)
	Same	Same	Same	Sperm penetration	Kraemer (1966)
	Follicular, matured in vitro	Semen	Modified Ham's F-10	2 polar bodies and cleavage	Dahlhausen et al. (1980)
	Oviductal	Ejaculated, preincubated	SOF	Sperm penetration and cleavage	Bondioli and Wright (1980)

embryo was transferred into the uterus during the early luteal phase of a natural menstrual cycle on one or several occasions (Edwards *et al.*, 1980; Lopata *et al.*, 1980). Three other women conceived but later spontaneously aborted. Two of the aborted fetuses had chromosomal aberrations; one, aborted at 7 weeks, was a 69XXX triploid fetus, and the second, aborted at 21 weeks, was an anatomically normal male fetus whose karyotype showed clear pleomorphism at 15D and a large Y chromosome. The third miscarriage involved a completely normal male fetus at 20 weeks following chorioamnionitis caused by an anaerobic gram-negative bacillus. Additional animal research, especially with nonhuman primates and other species resembling man in normal production of a single offspring with long gestation periods is indicated. The cow represents an ideal research subject (Brackett, 1978b) both for knowledge basic to an important human problem and, perhaps more important to society, for advancing breeding technology in a major food-producing animal.

D. *In Vitro* Fertilization in Domestic Animals

Among companion animals, fertilization has been achieved *in vitro* in the cat (Hamner *et al.*, 1970; Bowen, 1977) and the dog (Mahi and Yanagimachi, 1976), but routine procedures involving subsequent embryo transfer have not yet been developed. Several attempts have been made to fertilize ova of large domestic animals *in vitro* (for review, see Brackett, 1979b). Most of these reports (Table II) are preliminary and are best thought of as experience upon which to build. Conditions are not yet adequately defined to enable consistent and repeatable success with *in vitro* fertilization in large animals other than the cow.

The state-of-the-art of *in vitro* maturation of oocytes from ovarian follicles is not yet sufficiently developed so that viable embryos can be obtained via *in vitro* fertilization. It is now realized that cytoplasmic maturation of the oocyte is equally as important as the more easily observable and experimentally obtainable chromosomal maturation, i.e., meiotic progression to the arrested Metaphase II stage. The importance of the follicle in maintaining the appropriate environment for the developing ovum to become competent is now recognized. This role of the follicle is probably achieved *in vivo* by a classical sequential steroid action provoked by the dramatic influence of the LH surge (Thibault, 1977). In contrast to ooplasm, maturational changes of the zona pellucida are not recognized. There is some evidence in the rabbit that capacitated spermatozoa can penetrate the zonae of mature or immature oocytes with equal facility (Overstreet and Bedford, 1974). Since the conditioning of sperm cells to enable penetration of the zona pellucida is of special interest, immature or *in vitro* matured oocytes can be useful in assessing zona-penetrating ability of sperm cells.

The fertilizing ability of bull sperm observed with zona-free hamster ova was verified by penetration of zonae pellucidae of *in vitro* matured cow oocytes

(Brackett *et al.*, 1980b). A sperm tail within an ovum from one of these experiments can be seen in Fig. 1. In additional experiments, *in vitro* insemination of preovulatory and recently ovulated cow ova resulted in pronuclear, 2-cell, and 4-cell stages, and, with embryo transfer, normal gestational development (Brackett *et al.*, 1981). A cow embryo that developed normally in culture to the 4-cell stage after *in vitro* fertilization can be seen in Fig. 2 (Brackett *et al.*, 1980c). It is anticipated that the recent cow experiments will be repeatable and serve to facilitate the production of viable bovine embryos by fertilization *in vitro*.

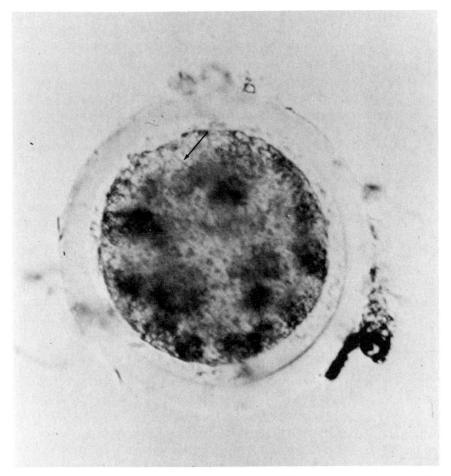

Fig. 1.   Cow oocyte penetrated by a bull sperm *in vitro* (525 ×). Sperm tail can be seen within the ooplasm (arrow). The oocyte was matured *in vitro* prior to *in vitro* insemination with *in vitro* capacitated bull sperm. (From Brackett *et al.*, 1980b by permission. Copyright 1980 by Elsevier North-Holland, Inc.)

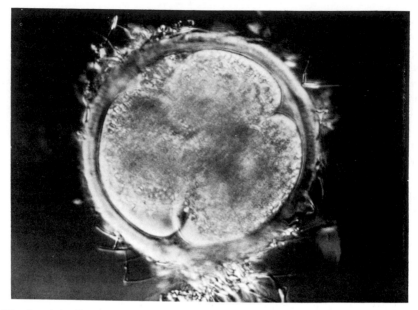

Fig. 2. A 4-cell stage cow embryo fertilized *in vitro* (400 ×). (From Brackett *et al.*, 1980c by permission. Copyright 1980 by Society for the Study of Reproduction.)

III. ASSESSMENT OF GAMETE FERTILIZING ABILITY

With *in vitro* fertilization, the treatment of one type gamete can be kept constant while experimentally manipulating influences upon the other to gain insight regarding effects on fertilizing ability. Thus, by using rabbit spermatozoa, capacitated *in vivo*, it was possible to demonstrate a superior fertilizing ability of oocytes recovered from the ovarian surface (Brackett and Server, 1970; Seitz *et al.*, 1970) followed in decreasing order by those recovered from oviducts (Brackett and Williams, 1968; Brackett, 1970; Harrison and Dukelow, 1972) and finally those from ovarian follicles (Brackett *et al.*, 1972; Mills *et al.*, 1973). Under defined *in vitro* conditions, ejaculated rabbit spermatozoa reflect individuality of bucks regarding relative effectiveness for fertilization of ova (Brackett and Oliphant, 1975). Also, epididymal sperm in the rabbit have been shown to undergo capacitation *in vitro* more easily and to fertilize higher proportions of ova than could ejaculated sperm from the same animals (Brackett *et al.*, 1978a). Genetically determined variability in combinations of gametes, mentioned above, is suspected but yet undefined in species other than the mouse.

A very practical application of *in vitro* fertilization techniques lies in the qualitative and quantitative assessment of sperm fertilizing ability (Brackett, 1979a). Thus, *in vitro* fertilization was used for testing of anti-fertility effects of

α-chlorohydrin on sperm fertilizing ability of treated rats and mice (Tsunoda and Chang, 1976). In these species, there is little advantage in use of the *in vitro* approach over more traditional *in vivo* fertilization tests since animals are usually sacrificed in each instance. However, in primate or large animal species the *in vitro* approach becomes more attractive. It is often possible to use valuable animals repeatedly as gamete donors.

In addition to the potential for understanding infertility, further development of this technology could be extremely important for determining the fertilizing ability of any sperm remaining in oligospermic ejaculates of men in trials concerned with male contraceptive development. Such information could make possible the avoidance of exposure of the female partner to fertile sperm cells that might, due to some derangement caused by the treatment, represent the risk of an abnormal pregnancy. Assessment of domestic animal sperm as an adjunct to improved breeding provides additional motivation for *in vitro* fertilization research. More appropriate handling of infertility cases (human or animal) as well as affording a means for contrasting fertilizing ability between two males is an obvious avenue for exploitation of this technology.

Both the ability to discern fertile from infertile spermatozoa and the ability to rank individual males according to the fertility of sperm cells they produce are potentially valuable. Simple definitive *in vitro* tests would find immediate application in early testing of "freezability" of bull semen. In current practice, bulls whose semen cannot withstand the rigors of freezing and thawing might be eliminated much earlier and at great savings. Such tests might also offer an attractive basis for inclusion of fertility in genetic selection. Greater fertility is sometimes more economically attractive than genetic merit based on production.

A. Exposure of Intact Ova to Spermatozoa

Several investigations have pointed to differences in ability of sperm samples to interact with intact ova. In one study, spermatozoa from fertile men interacted with zonae pellucidae of cadaver oocytes to a greater extent than did sperm from infertile men (Overstreet and Hembree, 1976). Heterospermic insemination of cows, i.e., artificial insemination with semen from two males which produce offspring that can be distinguished by coat color or blood type, required less than 1/70 the number of inseminations for statistical evaluation of superior fertilizing sires compared with homospermic inseminations (Beatty *et al.*, 1969). Direct and rapid comparisons might become possible by heterospermic or competitive fertilization *in vitro*. With fluorescent marking of one sperm sample, it was apparent from preliminary rabbit work that epididymal sperm, which fertilize higher proportions of ova *in vitro* (Brackett *et al.*, 1978a), were associated to a greater extent with ova *in vitro* than were ejaculated sperm treated similarly to effect capacitation (Brackett, 1979a).

An investigation of idiopathic infertility in human couples involved *in vitro* insemination of mature oocytes obtained at diagnostic laparoscopy after clomiphene and human chorionic gonadotropin treatment (Trounson *et al.*, 1980b). Encouragement for assessing fertility was provided by the observations that apparently normal embryos developed from five of six oocytes from women with blocked oviducts, whereas no normal embryos developed from oocytes from nine patients who had sustained infertility of unknown cause for 2 years or more. The latter group exhibited a high degree of fertilization failure and severe polyspermy occurred in two cases. These findings point to the appropriateness of additional studies (Trounson *et al.*, 1980b). In case of fertilization failure, sera and follicular fluids should be examined for the presence of antizona antibodies shown to reduce sperm binding to the oocyte and fertilization *in vitro* (Trounson *et al.*, 1980a) and suggested from many animal studies to be a promising immunological approach to contraception (Shivers and Sieg, 1980). Also, spermatozoa should be examined by electron microscopy for morphologic abnormalities (Sun and White, 1978); additional studies of fertilizing capacity of the sperm in question can be further explored by sperm penetration tests of salt-stored human zonae (Yanagimachi *et al.*, 1979). Obviously, the utility of the *in vitro* approach for assessing sperm fertilizing ability must be based on extensive studies correlating both *in vivo* and *in vitro* investigation. The latter consideration emphasizes the importance of development of this methodology in domestic animals since ample data on conception rates can be attributed to individual sires.

Attachment to and penetration of the zona pellucida by spermatozoa is species specific except when the species are closely related (e.g., sheep and goat). It has been suggested that research involving gamete interaction might be useful in efforts to better understand man's evolutionary origin (Short, 1979a; Soupart, 1979a, b). Human sperm were found to attach to and penetrate the zona pellucida of the gibbon oocyte *in vivo* and *in vitro* (Bedford, 1977). It seems unlikely that oocytes from endangered species of apes could be available in adequate supply for testing the zona-penetrating ability of human sperm. However, there are many occasions where gamete interactions of closely related but different species might be useful, e.g., endangered species maintained by zoological societies and their domesticated "cousins."

B. Studies with Zona-Free Hamster Ova

The zona-free hamster ovum test was applied initially to human sperm (Yanagimachi *et al.*, 1976) and efforts are underway in several laboratories to correlate fertile and "suspect" human sperm with results of this test (Barros *et al.*, 1979; Hall *et al.*, 1979; Rogers *et al.*, 1979; Binor *et al.*, 1980). The negative results are held to be most meaningful since lack of ability to effect capacitation and the acrosome reaction experimentally suggests an inherent defect of the sperm population. Insight into unexplained infertility followed observation of differences in degrees of human sperm performance when immature oocytes were used for assessing zona penetration along with zona-free hamster ova (Overstreet *et al.*, 1980). Further, it was suggested that development of this methodology might become useful in identifying couples with male infertility as candidates for *in vitro* fertilization and embryo transfer. Efforts have also been initiated to employ zona-free hamster ovum methodology for examining the chromosome complements of human sperm cells (Rudak *et al.*, 1978). The zona-free hamster ovum test no longer requires maintenance of a hamster colony since fertilizability of these preparations has been demonstrated following cryopreservation (Fleming *et al.*, 1979).

Efforts to capacitate freshly ejaculated bull sperm under defined conditions have led to consistent results in that high proportions of zona-free hamster ova contained male pronuclei by 6 hr after insemination with sperm from different bulls (Brackett *et al.*, 1980b). Distinct differences in sperm penetration and pronuclear development within zona-free hamster ova by frozen-stored semen from two bulls of high fertility were demonstrable (Bousquet and Brackett, 1981). More research is required to establish the putative utility of this approach.

IV. TREATMENT OF INFERTILITY

A. Human Studies

Success in overcoming infertility attributed to blocked oviducts has been reported (Table I) following recovery of preovulatory follicular ova from women with normal menstrual cycles, coupled with *in vitro* fertilization and embryo transfer (Steptoe and Edwards, 1978; Edwards *et al.*, 1980; Lopata *et al.*, 1980; Steptoe *et al.*, 1980). In addition to selected infertility cases involving blocked oviducts and oligospermia, Soupart has emphasized the potential use of *in vitro* fertilization and embryo culture combined with karyotyping in cases of high genetic risk to prevent birth defects (Soupart, 1979a, b). The latter application is obviously for the more distant future since it would require improved methods for culture and biopsy of embryos. Lopata (1980) summarized current human embryo transfer efficiency and the probability of having a child by *in vitro* fertilization and embryo transfer; three live births resulted from 56 embryos transferred, 5.4% successful embryo transfer. However, a more realistic perspective follows when data are examined according to total number of subjects (treatment cycles) participating in the British and Australian *in vitro* fertilization and embryo transfer programs. Of 210 women, six became pregnant and of these, three delivered normal infants, 1.43% success (or a 98.57% disappointment rate among hopeful, infertile patients). Technologies required for human *in vitro* fertilization are rapidly advancing; these include endocrine monitoring, oocyte recovery after hormonal treatments, *in vitro* fertilization with culture, and transfer into the appropriately synchronous uterine environment. The development of superovulation with *in vitro* fertilization and freezing of human embryos may someday make subsequent transfer possible without oocyte recovery and *in vitro* fertilization procedures each time. Ethical concerns are inherent in human *in vitro* fertilization and much attention has surrounded these important technologies and their applications in human medicine and will continue to do so.

B. Animal Infertility and *in Vitro* Fertilization

Infertility in the cow, mare, or other female mammal may reflect disease conditions (1) that result in defective egg production, (2) that prevent fertilization of a normal egg, or (3) that prevent development of the embryo following fertilization. In the first instance the ovary and/or ovulatory mechanism would be at fault, whereas in the second and third instances, the female reproductive tract might be involved with varying degrees of pathology, or the problem might be attributable to the male, e.g., oligospermia. Such problems, then, result in fertilization failure or embryonic mortality. In a study of repeat breeder cows, 40.8% fertilization failure was observed (Tanabe and Casida, 1949). If viable ova could be recovered from such cows, *in vitro* fertilization should provide the means for overcoming problems of sperm–egg meeting. The use of *in vitro* fertilization to overcome oligospermia might be useful in horses and certain valuable bulls (or other males) where a genetic cause of the problem can be ruled out.

An unfavorable milieu for embryonic development might be avoided if a viable embryo can be obtained and transferred to a healthy recipient. Indeed, embryo transfer has already been employed to overcome such reproductive problems (Bowen *et al.*, 1978). Overall, 14 (56%) of 25 selected cows were successfully treated by embryo transfer to overcome subnormal fertility. If the clinical armamentarium also included *in vitro* fertilization, even greater success rates might ultimately be achieved.

V. COMBINATION OF SELECTED GAMETES FOR ANIMAL PRODUCTION

A. Alternatives to *in Vitro* Fertilization

The vulnerability of preovulatory or recently ovulated mammalian ova, along with the difficulties associated with sperm preparation for fertilization *in vitro*, has prompted research toward achieving early embryonic development within the reproductive tract of a different animal species. In earlier work, zygotes and/or early cleavage stage embryos of several species (pig, sheep, horse, cow) have been observed to continue development and retain their viability following interim transfer to the rabbit oviduct (see Chapter 3, this volume). The pig oviduct can support the penetration of cow ova by bull sperm (Bedirian *et al.*, 1975). The oviductal environment of pseudopregnant rabbit does has been found capable of supporting fertilization of the squirrel monkey and of the golden hamster, a process referred to as *xenogenous* fertilization (DeMayo *et al.*, 1980). Recently, xenogenous fertilization was accomplished using bovine, porcine, and hamster follicular oocytes inseminated within oviducts of pseudopregnant rabbits (Hirst *et al.*, 1981). A combination of *in vitro* and xenogenous fertilization methodologies has recently led to cow embryonic development to the morula stage during 4 days within the rabbit female reproductive tract (M. B. Wheeler, R. P. Elsden, and G. E. Seidel, Jr., personal communication). This approach might prove useful for many of the same applications as *in vitro* fertilization. Furthermore, these exciting observations emphasize the importance of comparative studies and the potential utility of the rabbit as a model for development of practical technologies for animal breeding.

Other alternatives to *in vitro* insemination for combining selected gametes to result in genetically desired embryonic combinations include technologies employing micromanipulation and microsurgery (see Chapter 10, this volume). The fertilization process might be initiated through microinjection of sperm cells directly into the ovum (Lin, 1971). Male pronuclear development within hamster ova occurred after microinjection of hamster sperm nuclei or nuclei of fresh, frozen–thawed, or freeze-dried human sperm (Uehara and Yanagimachi, 1976). The latter work indicates that sperm nuclei are quite stable and that pronuclear development within the ooplasm is not strictly species specific. Although not promising at present, developments of such *in vitro* fertilization technologies in the broad sense might be reasonably anticipated for future applications in domestic animal breeding.

B. Fertilization *in Vitro*

Already, *in vitro* fertilization and embryo transfer potentially provide a major advantage in enabling development from a single superovulation procedure of several embryos with different sires. This and other advantages will be magnified with fruition of current efforts to develop procedures for frozen storage of unfertilized eggs. Successful *in vitro* fertilization has been reported after freezing and thawing of oocytes from the mouse (Parkening *et al.*, 1976; Tsunoda *et al.*, 1976; Parkening and Chang, 1977; Whittingham, 1977), hamster (Tsunoda *et al.*, 1976), and rat (Kasai *et al.*, 1978). With further development of *in vitro* fertilization methodology along with the facility to store unfertilized oocytes, gamete banking and fertilization of desired crosses should become possible. Methods for separation of X- and Y-chromosome containing sperm cells might become possible only when few sperm cells are involved, too few for successful use in artificial insemination. Then, *in vitro* fertilization might allow the initiation of development of individuals of the desired sex, as well. An additional advantage would be the facility afforded by this approach to spread desirable genetic influences of a top bull (or other male) much further by bringing a few of his sperm cells into contact with each ovum destined, following embryo transfer, to develop into a superior offspring. Although efforts toward genetic engineering at the level of germ cells is most obviously handled through egg treatments, the development of *in vitro* procedures for introducing the desired DNA into embryos through the sperm cell at fertilization (Brackett *et al.*, 1971) might someday be very useful. With long-term embryo storage, the resulting embryos might be stored or transferred immediately into recipients for term development. Indeed, as is the case for many other new technologies in animal breeding, *in vitro* fertilization can become more useful when combined with other reproductive technologies receiving attention of research scientists today. Rapid technological developments are anticipated within the next two decades. The species in which application of such technology is emphasized will depend on sociological, economic, and political determinants.

ACKNOWLEDGMENTS

The author thanks George E. Seidel, Jr., and Sarah M. Seidel for critically reading the manuscript and Pamela J. Salsbury for typing it. Research support of Grant HD09406 from the National Institute of Child Health and Human Development, National Institutes of Health, is gratefully acknowledged.

REFERENCES

Austin, C. R. (1974). *In* "MTP International Review of Science, Reproductive Physiology" (R. O. Greep, ed.), pp. 95–131. Univ. Park Press, Baltimore, Maryland.
Barros, C., and Yanagimachi, R. (1972). *J. Exp. Zool.* **180**, 251.
Barros, C., Gonzalez, J., Herrera, E., and Bustos-Obregon, E. (1979). *Andrologia* **11**, 197–210.
Bavister, B. D. (1980). *Dev. Growth Differ.* **22**, 385–402.
Bavister, B. D. (1981). *In* "Fertilization and Embryonic Development *In Vitro*" (L. Mastroianni, Jr., and J. Biggers, eds.), pp. 42–60. Plenum, New York.
Beatty, R. A., Bennett, J. H., Hall, J. G., Hancock, J. L., and Stewart, D. L. (1969). *J. Reprod. Fertil.* **19**, 491–502.
Bedford, J. M. (1977). *Anat. Rec.* **188**, 477–488.
Bedford, J. M. and Chang, M. C. (1962). *Nature (London)* **193**, 808–809.
Bedirian, K. N., Shea, B. F., and Baker, R. D. (1975). *Can. J. Anim. Sci.* **55**, 251–256.

Biggers, J. D. (1981). *N. E. J. Med.* **304,** 336–342.

Binor, Z., Sokoloski, J. E., and Wolf, D. P. (1980). *Fertil. Steril.* **33,** 321–327.

Bondioli, K. R., and Wright, R. W., Jr. (1980). *J. Anim. Sci.* **51,** 660–667.

Bousquet, D., and Brackett, B. G. (1981). *Theriogenology* **15,** 117.

Bowen, R. A. (1977). *Biol. Reprod.* **17,** 144–147.

Bowen, R. A., Elsden, R. P., and Seidel, G. E., Jr. (1978). *J. Am. Vet. Med. Assoc.* **172,** 1303–1306.

Brackett, B. G. (1969). *Fertil. Steril.* **20,** 127–142.

Brackett, B. G. (1970). *Fertil. Steril.* **21,** 169–176.

Brackett, B. G. (1978a). *Environ. Health Perspect.* **24,** 65–71.

Brackett, B. G. (1978b). Hearing before the Subcommittee on Health and the Environment of the Committee on Interstate and Foreign Commerce, Ninety-Fifth Congress, Serial No. 95-134. U.S. Government Printing Office, Washington, D.C.

Brackett, B. G. (1979a). *In* "Animal Models for Research on Contraception and Fertility" (N. J. Alexander, ed.), pp. 254–268. Harper & Row, Hagerstown, Maryland.

Brackett, B. G. (1979b). *Beltsville Symp. Agric. Res.* **3,** 171–193.

Brackett, B. G. (1980). *In* "Gynecologic Endocrinology" (J. J. Gold, and J. B. Josimovich, eds.), pp. 890–924. Harper & Row, Hagerstown, Maryland.

Brackett, B. G. (1981). *In* "Fertilization and Embryonic Development *In Vitro*" (L. Mastroianni, Jr., and J. Biggers, eds.), pp. 61–79. Plenum Press, New York.

Brackett, B. G., and Oliphant, G. (1975). *Biol. Reprod.* **12,** 260–274.

Brackett, B. G., and Server, J. B. (1970). *Fertil. Steril.* **21,** 687–695.

Brackett, B. G., and Williams, W. L. (1968). *Fertil. Steril.* **19,** 144–155.

Brackett, B. G., Baranska, W., Sawicki, W., and Koprowski, H. (1971). *Proc. Natl. Acad. Sci. U.S.A.* **68,** 353–357.

Brackett, B. G., Mills, J. A., and Jeitles, G. G., Jr. (1972). *Fertil. Steril.* **23,** 898–909.

Brackett, B. G., Oh, Y. K., Evans, J. F., and Donawick, W. J. (1977). *10th Annu. Meet. Soc. Study Reprod.,* pp. 56. (Abstr. 86).

Brackett, B. G., Hall, J. L., and Oh, Y. K. (1978a). *Fertil. Steril.* **29,** 571–582.

Brackett, B. G., Oh, Y. K., Evans, J. F., and Donawick, W. J. (1978b). *Theriogenology* **9,** 89.

Brackett, B. G., Evans, J. F., and Donawick, W. J. (1980a). *9th Int. Congr. Anim. Reprod. Artif. Insem.,* Madrid, **3,** 296.

Brackett, B. G., Evans, J. F., Donawick, W. J., Boice, M. L., and Cofone, M. A. (1980b). *Arch. Androl.* **5,** 69–71.

Brackett, B. G., Oh, Y. K., Evans, J. F., and Donawick, W. J. (1980c). *Biol. Reprod.* **23,** 189–205.

Brackett, B. G., Bousquet, D., Boice, M. L., Donawick, W. J., and Evans, J. F. (1981). *Biol. Reprod.* **24,** *Suppl. 1,* 173, 109A.

Bregulla, K., Gerlach, U., and Hahn, R. (1974). *Dtsch. Tieraerztl. Wochenschr.* **81,** 465–470.

Chang, M. C. (1959). *Nature (London)* **184,** 466–467.

Chang, M. C., Austin, C. R., Bedford, J. M., Brackett, B. G., Hunter, R. H. F., and Yanagimachi, R. (1977). *In* "Frontiers in Reproduction and Fertility Control" (R. O. Greep, and M. A. Koblinsky, eds.), pp. 434–451. MIT Press, Cambridge, Massachusetts.

Cross, P. C., and Brinster, R. L. (1970). *Biol. Reprod.* **3,** 298–307.

Dahlhausen, R. D., Dresser, B. L., and Ludwick, T. M. (1980). *Theriogenology* **13,** 93.

Dauzier, L., and Thibault, C. (1959). *C. R. Acad. Sci.* **248,** 2655–2656.

DeMayo, F. J., Mizoguchi, H., and Dukelow, W. R. (1980). *Science (Washington, D.C.)* **208,** 1468–1469.

Dukelow, W. R. (1978). *In* "Methods in Mammalian Reproduction" (J. C. Daniel, Jr., ed.), pp. 438–458. Academic Press, New York.

Edwards, R. G., Steptoe, P. C., and Purdy, J. M. (1980). *Br. J. Obstet. Gynaecol.* **87,** 737–756.

Fleming, A. D., Yanagimachi, R., and Yanagimachi, H. (1979). *Gam. Res.* **2**, 357–366.

Fraser, L. R. (1979). *J. Reprod. Fertil.* **55**, 153–160.

Fraser, L. R., and Dandekar, P. V. (1973a). *J. Exp. Zool.* **184**, 303–312.

Fraser, L. R., and Dandekar, P. V. (1973b). *J. Reprod. Fertil.* **33**, 159–161.

Fraser, L. R., and Drury, L. M. (1975). *Biol. Reprod.* **13**, 513–518.

Fraser, L. R., Dandekar, P. V., and Vaidya, R. A. (1971). *Biol. Reprod.* **4**, 229–233.

Gwatkin, R. B. L. (1977). "Fertilization Mechanisms in Man and Mammals." Plenum, New York.

Hall, J. L. (1981). *Ob/Gyn Survey* (submitted).

Hall, J. L., and Goulding, E. (1981). *Fertil. Steril.* (submitted).

Hall, J. L., Sloan, C. S., and Willis, W. D. (1979). *Biol. Reprod.* **20**, *Suppl. 1*, 169, 96A.

Hamner, C. E., Jennings, L. L., and Sojka, N. J. (1970). *J. Reprod. Fertil.* **23**, 477–480.

Harrison, R. M., and Dukelow, W. R. (1972). *J. Reprod. Fertil.* **31**, 483–486.

Hirst, P. J., DeMayo, F. J., and Dukelow, W. R. (1981). *Theriogenology* **15**, 67–75.

Hoppe, P. C., and Pitts, S. (1973). *Biol. Reprod.* **8**, 420–426.

Imai, H., Niwa, K., and Iritani, A. (1977). *J. Reprod. Fertil.* **51**, 495–497.

Imai, H., Niwa, K., and Iritani, A. (1979). *J. Reprod. Fertil.* **56**, 489–492.

Iritani, A., and Niwa, K. (1977). *J. Reprod. Fertil.* **50**, 119–212.

Iritani, A., Sato, E., and Nishikawa, Y. (1974). *Soc. Study Reprod. Annu. Meet.* (Abstr.), pp. 115–116.

Jeitles, G. G., Jr., and Brackett, B. G. (1974). *Fed. Proc. Fed. Am. Soc. Exp. Biol.* **33**, 381 (Abstr. 440).

Kajii, T., and Oknma, K. (1977). *Nature (London)* **268**, 633–634.

Kasai, M., Niwa, K., and Iritani, A. (1978). *Cryobiology* **15**, 680.

Kaufman, M. H. (1973). Ph.D. Thesis, University of Cambridge, England.

Kaufman, M. H., and Whittingham, D. G. (1972). *J. Reprod. Fertil.* **28**, 465–468

Kraemer, D. C. (1966). *Diss. Abstr.* **27**, 2858.

Lin, T. P. (1971). *In* "Methods in Mammalian Embryology" (J. C. Daniel, Jr., ed.), pp. 157–171. Academic Press, New York.

Lopata, A. (1980). *Nature (London)* **288**, 642–643.

Lopata, A., Johnston, I. W. H., Hoult, I. J., and Speirs, A. I. (1980). *Fertil. Steril.* **33**, 117–120.

Lorton, S. P., and First, N. L. (1979). *Biol. Reprod.* **21**, 301–308.

Mahi, C. A., and Yanagimachi, R. (1976). *J. Exp. Zool.* **196**, 189–196.

McRorie, R. A., and Williams, W. L. (1974). *Annu. Rev. Biochem.* **43**, 777–804.

Meizel, S. (1978). *Dev. Mamm.* **3**, 1–64.

Mills, J. A., Jeitles, G. G., Jr., and Brackett, B. G. (1973). *Fertil. Steril.* **24**, 602–608.

Miyamoto, H., and Chang, M. C. (1972). *J. Reprod. Fertil.* **30**, 135–137.

Mukherjee, A. B. (1972). *Nature (London)* **237**, 397–398.

Mukherjee, A. B., and Cohen, N. M. (1970). *Nature (London)* **228**, 472–473.

Mullen, R. J., and Carter, S. C. (1973). *Biol. Reprod.* **9**, 111–115.

Oliphant, G., and Brackett, B. G. (1973). *Biol. Reprod.* **9**, 404–414.

Overstreet, J. W., and Bedford, J. M. (1974). *Dev. Biol.* **41**, 185–191.

Overstreet, J. W., and Hembree, W. C. (1976). *Fertil. Steril.* **27**, 815–831.

Overstreet, J. W., Yanagimachi, R., Katz, D. F., Hayaski, K., and Hanson, F. W. (1980). *Fertil. Steril.* **33**, 534–542.

Parkening, T. A., and Chang, M. C. (1976a). *J. Reprod. Fertil.* **48**, 381–383.

Parkening, T. A., and Chang, M. C. (1976b). *Biol. Reprod.* **15**, 647–653.

Parkening, T. A., and Chang, M. C. (1977). *Biol. Reprod.* **17**, 527–531.

Parkening, T. A., Tsunoda, Y., and Chang, M. C. (1976). *J. Exp. Zool.* **197**, 369–374.

Rogers, B. J. (1978). *Gam. Res.* **1**, 165–223.

Rogers, B. J., Van Campen, H., Ueno, M., Lambert, H., Bronson, R., and Hale, R. (1979). *Fertil. Steril.* **32**, 664–670.

Rudak, E., Jacobs, P. A., and Yanagimachi, R. (1978). *Nature (London)* **274**, 911–913.

Schneider, U., Hahn, J., and Sulzer, H. (1974). *Dtsch. Tieraerztl. Wochenschr.* **81**, 470.

Seidel, G. E. Jr., Bowen, R. A., and Kane, M. T. (1976). *Fertil. Steril.* **27**, 861–870.

Seitz, H. H., Jr., Brackett, B. G., and Mastroianni, L., Jr. (1970). *Biol. Reprod.* **2**, 262–267.

Shivers, C. A., and Seig, P. M. (1980). *In* "Immunological Aspects of Infertility and Fertility Regulation" (D. S. Dhindsa and G. F. B. Schumacher, eds.), pp. 173–182. Elsevier North-Holland, Amsterdam.

Short, R. V. (1979a). Study #10, Ethics Advisory Board, Dept. of HEW. Appendix: HEW Support of research involving human in vitro fertilization and embryo transfer. U.S. Govt. Printing Office, Washington, D.C. (Stock No. 017-040-00454-1).

Short, R. V. (1979b). *In* "Maternal Recognition of Pregnancy," Ciba Foundation Symposium 64 (new series), pp. 377–387. Excerpta Medica, Amsterdam.

Soupart, P. (1979a). *Curr. Probl. Obstet. Gynecol.* **3** (No. 2, Part I), 5–45.

Soupart, P. (1979b). *Curr. Probl. Obstet. Gynecol.* **3** (No. 3, Part II), 5–43.

Steptoe, P. C., and Edwards, R. G. (1970). *Lancet* **1**, 683–689.

Steptoe, P. C., and Edwards, R. G. (1978). *Lancet* **2**, 366.

Steptoe, P. C., Edwards, R. G., and Purdy, J. M. (1980). *Br. J. Obstet. Gynecol.* **87**, 757–768.

Sun, C. N., and White, H. J. (1978). *Cytologia* **43**, 551.

Tanabe, T. Y., and Casida, L. E. (1949). *J. Dairy Sci.* **32**, 237–246.

Thibault, C. (1977). *J. Reprod. Fertil.* **51**, 1–15.

Thibault, C., and Dauzier, L. (1961). *Ann. Biol. Anim. Biochim. Biophys.* **1**, 277–294.

Thibault, C., Gerard, M., and Menezo, Y. (1975). *Ann. Biol. Anim. Biochem. Biophys.* **15**, 705–715.

Toyoda, Y., and Chang, M. C. (1974). *J. Reprod. Fertil.* **36**, 9–22.

Trounson, A. O., Shivers, C. A., McMaster, R., and Lopata, A. (1980a). *Arch. Androl.* **4**, 29–36.

Trounson, A. O., Leeton, J. F., Wood, C., Webb, J., and Kovacs, G. (1980b). *Fertil. Steril.* **34**, 431–438.

Tsunoda, Y., and Chang, M. C. (1976). *J. Reprod. Fertil.* **46**, 401–406.

Tsunoda, Y., Parkening, T. A., and Chang, M. C. (1976). *Experientia* **32**, 223–224.

Uehara, T., and Yanagimachi, R. (1976). *Biol. Reprod.* **15**, 467–470.

Whittingham, D. G. (1968). *Nature (London)* **220**, 592–593.

Whittingham, D. G. (1975) *In* "The Early Development of Mammals" (M. Balls and A. E. Wild, eds.), pp. 1–24. Cambridge Univ. Press, London and New York.

Whittingham, D. G. (1977). *J. Reprod. Fertil.* **49**, 89–94.

Whittingham, D. G. (1979). *Br. Med. Bull.* **35**, 105–111.

Wilson, J. G. (1973). "Environment and Birth Defects." Academic Press, New York.

Yanagimachi, R. (1977) *In* "Immunobiology of Gametes" (M. Edidin and H. H. Johnson, eds.), pp. 255–295. Cambridge Univ. Press, London and New York.

Yanagimachi, R. (1981). *In* "Fertilization and Embryonic Development *in Vitro*" (L. Mastroianni, Jr., and J. Biggers, eds.), pp. 82–182. Plenum, New York.

Yanagimachi, R., Yanagimachi, H., and Rogers, B. J. (1976). *Biol. Reprod.* **15**, 471–476.

Yanagimachi, R., Lopata, A., Odom, A., Bronson, R. A., Mahi, C. A., and Nicholson, G. L. (1979). *Fertil. Steril.* **31**, 562–574.

9

Amphibian Nuclear Transplantation: State of the Art

ROBERT GILMORE McKINNELL

I. INTRODUCTION

The most rigorous test of gene content and function is that of nuclear transplantation (Briggs, 1977).

Amphibian nuclear transplantation, also known as *cloning*, is an experimental procedure that was *not* designed to enhance animal production. Cloning was designed as a procedure to ascertain whether or not somatic nuclei of developing embryos (and ultimately, somatic nuclei of adult organisms) retain all of the genetic information found in a zygote nucleus and whether or not a somatic

163

NEW TECHNOLOGIES IN ANIMAL BREEDING

nucleus will serve as a surrogate for the zygote nucleus. The procedure is a rigorous functional test of the entire genome of a living cell (DiBerardino, 1980; DiBerardino and Hoffner, 1980). Thus, nuclear transplantation is a bioassay procedure. It is not a technique designed to be of use to the animal breeder.

Despite the fact that animal cloning was not designed as a reproductive procedure, there has been considerable speculation concerning its potential benefits in animal production (e.g., see Chapters 6, 10, 11, and 12, this volume). It is easier to interpret the significance of that speculation with an understanding of the progress that has been made thus far with animal cloning. The emphasis will be on amphibian experiments because more laboratories have encountered success with amphibian cloning than with other animal groups. Methods and results of experimental cloning studies will be reviewed. The rationale for the investigations will be presented. Cloning successes, which resulted in normal animals, will be discussed and contrasted with transplantation studies of adult nuclei that failed to yield normal embryos. Finally, a brief account of progress with the cloning of small mammalian species will be considered.

II. METHODOLOGY

It is possible to make two generalizations concerning the methodology of amphibian nuclear transplantation. The first concerns the fact that although the manipulations attendant to cloning require extraordinary skill (Briggs, 1977, described it as "a rather difficult technique"), the procedure is conceptually simple. Amphibian cloning requires only the insertion of a totipotent nucleus into an enucleated mature ovum. Few biologically significant scientific procedures can be described as simply. The accomplishment of this procedure, however, requires unusual microsurgical skill. It also requires the guaranteed availability of healthy frogs for the production of eggs and embryos when needed. The second generalization concerns the remarkable fact that the technology of cloning has remained virtually unchanged since its beginnings (Briggs and King, 1952). The first species of amphibian to be cloned was the Northern Leopard Frog, *Rana pipiens*. Most of my experience has been with *R. pipiens*. Accordingly, methodology described in this review relates primarily to that anuran. However, only minor changes in technique are required to extend the manipulative procedures to other species.

A. Preparation of Recipient Ovum

Eggs are obtained in season (October through June) from a mature female *R. pipiens* that has been injected with pituitary glands (Rugh, 1934) or pituitary glands and progesterone (Wright and Flathers, 1961). Forty-eight hours after treatment, freshly ovulated eggs are extruded onto a glass slide, covered with a dilute electrolyte solution, and activated with the prick of a clean glass needle. The needle prick will set into motion the meiotic maturation events that would cause the formation of a second polar body if enucleation did not take place. The events permit enucleation with ease.

It is absolutely imperative that the recipient ovum be properly and reliably enucleated. Maternal chromosomes, when retained in the egg that receives a transplanted nucleus, are competent to fuse with the inserted nucleus and participate in subsequent development (Sambuichi, 1959; Subtelny and Bradt, 1960, 1963; Kawamura and Nishioka, 1963; McKinnell, 1964). Because the cloner seeks genetic expression of the inserted nucleus alone (i.e., the egg nucleus should not participate in development with the inserted nucleus), enucleation efficacy cannot be stressed too much. Fortunately, a variety of procedures are available that provide assurance of reliable nuclear extirpation without permanent damage to the ovum.

Amphibian ova are large (many are in the range of 1-mm diameter) and the cytoplasm is opaque. How then can the microscopic meiotic spindle with its maternal chromosomes be detected and ablated? Freshly extruded ova of both *Rana* and *Xenopus* are at metaphase of meiosis II. The second meiotic division, with the production of polar body II, occurs *after* insemination or artificial activation. A needle prick activates the egg so that rotation occurs and the dark animal hemisphere comes to lie uppermost. The meiotic apparatus is near the center of the animal hemisphere.

The polar body obviously cannot be extruded if the meiotic spindle were not adjacent to the egg surface. The proximity of the chromosomes to the surface of the egg results in the vulnerability of that maternal genetic material to a variety of experimental procedures. The position of the spindle is particularly easy to estimate in eggs of *R. pipiens* with the aid of a dissecting microscope. The meiotic apparatus, with melanin granules clustered about it, is located in a depression on the surface of the egg and appears as a distinct black dot. The black dot can be extirpated surgically (Porter, 1939), or it can be physiologically ablated with a pulse of ruby laser irradiation (McKinnell *et al.*, 1969b; Ellinger *et al.*, 1975). The laser enucleation procedure works equally well with ova of *Bombina orientalis* (Ellinger, 1979). The meiotic chromosomes of *Xenopus laevis* ova, as well as salamander ova, are effectively removed with ultraviolet irradiation (Gurdon, 1960; Briggs *et al.*, 1964).

A means of assuring that maternal chromosomal ablation was effective is provided by cytogenetic analysis of embryos that resulted from ova inseminated with untreated sperm prior to enucleation. It should be noted that ova are *not* inseminated prior to enucleation when they are to be used in a nuclear transfer experiment. The sperm nucleus in the inseminated ovum is protected from the effects of radiation by the yolky cytoplasm. Cytogenetically confirmed androgenetic haploid embryos result after radiation ablation of maternal chromosomes in both *R. pipiens* (McKinnell *et al.*, 1969b; Ellinger *et al.*, 1975) and *X. laevis* (R. G. McKinnell, M. Myers, B. T. Kren, and T. Byrne, unpublished observations) thereby demonstrating the effectiveness of the enucleation procedure. Diploid embryos ensue when enucleation does not occur (Fig. 1).

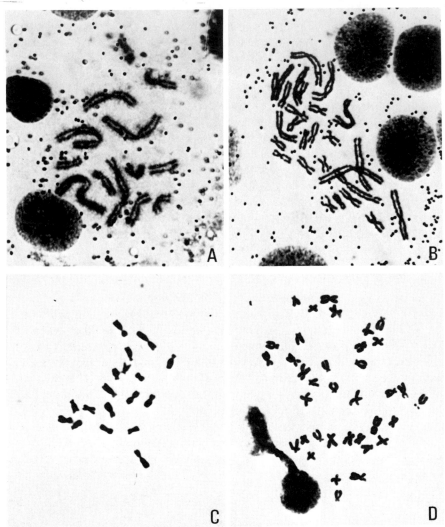

Fig. 1. Radiation ablation of maternal chromosomes results in haploid embryos of *R. pipiens* (A) and *X. laevis* (C). Unirradiated controls are diploid *R. pipiens* (B) and *X. laevis* (D). The radiation source used with *R. pipiens* was a ruby laser and that for *X. laevis* was ultraviolet [(A) and (B) above from Ellinger *et al.*, 1975].

Damage to the egg cortex undoubtedly results with any enucleation procedure. However, the damage is either so minimal that it does not affect the development of the clone, or the damage is quickly repaired. Regardless of which of these interpretations is correct, the significant observation is that normal development often occurs after inserting a totipotent somatic nucleus into an enucleated ovum.

Jelly removal is the next step in preparation of the recipient ovum. This can be accomplished either by cutting away the jelly with fine scissors and jewelers forceps or dissolution of the jelly with an enzyme preparation (Hennen, 1973). The activated, enucleated, and dejellied ovum is now ready to become the recipient of an inserted nucleus.

Before proceeding further, perhaps it would be appropriate to point out that eggs suitable for nuclear transplantation are not always available. Female *X. laevis* produce eggs with variable fertilizability. While some matings result in over 95% successful fertilization, the average fertility is about 50% (Gurdon, 1967). More recently, we reported a higher average fertilization rate of 71.9% (McKinnell *et al.*, 1981). Oocytes of *R. pipiens* vary in quality and fertilizability of eggs is reduced in some populations (McKinnell *et al.*, 1979); McKinnell and Schultheis, 1979). Hence, it is possible for a cloning experiment to be delayed, not owing to inadequate instruments or microsurgical skills, but because of failure to obtain high quality eggs for experimentation.

B. Preparation of Donor Cell for Transplantation

The biologically significant entity that is inserted into an activated enucleated recipient ovum in a cloning experiment is a nucleus. However, a cell with a broken plasma membrane is transplanted. There is no need to clean cytoplasm from the donor nucleus because such cleaning will damage the nucleus. Naked nuclei swell when placed in hypertonic solutions (Battin, 1959). Because nuclear damage occurs in artificial media, the donor cytoplasm is left relatively undisturbed to permit it to serve as a protective buffer between the nucleus and the transplantation medium.

Dissociation of young embryonic donor cells is accomplished with a calcium- and magnesium-free amphibian culture medium. Formulae for appropriate solutions are given elsewhere (McKinnell, 1978). Donor cells from older embryos, larvae, and adults require elevated pH, chelating agents, and/or proteolytic enzymes for proper dissociation (Moscona, 1952; Townes and Holtfreter, 1955; Jones and Elsdale, 1963).

C. Nuclear Insertion

A bevelled and sharpened micropipette is essential for nuclear transplantation. Glass capillary tubing, 1-mm diameter, is drawn to a tapering micropipette by

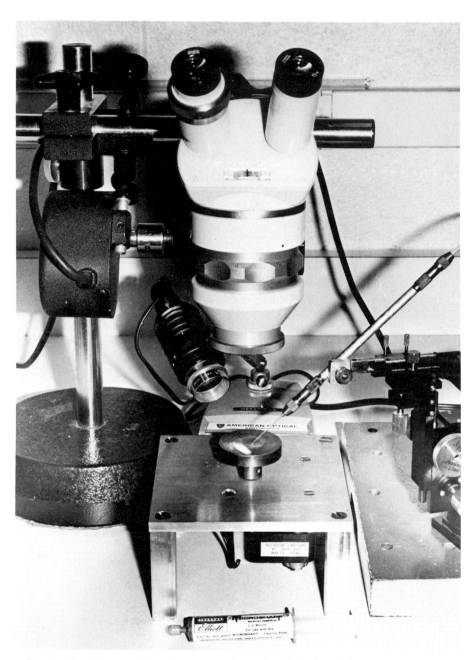

Fig. 2. Revolving turntable used with diamond paste for grinding bevelled tips on micropipettes.

any of several microneedle pullers commercially available (McKinnell, 1978). The micropipette tip can be fabricated entirely with aid of a microforge (McKinnell, 1978) or the bevel can be ground with diamond paste on a revolving turntable (Fig. 2) and the tip of the beveled micropipette (Fig. 3) sharpened with the microforge (Fig. 4).

The bevelled and sharpened micropipette is positioned with a microinjection apparatus held with a micromanipulator (Fig. 5). Fluid is drawn into or expelled from the micropipette with the microinjection apparatus.

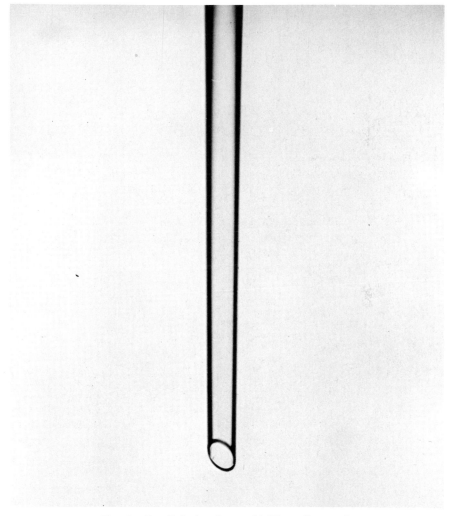

Fig. 3.   Bevelled micropipette with 37-μm-diameter tip.

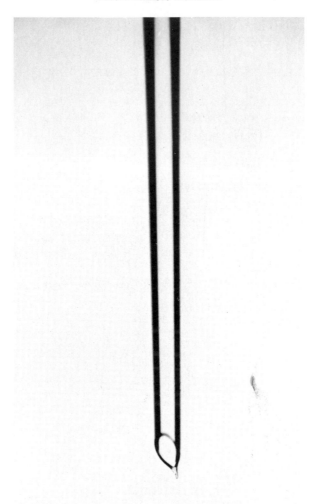

Fig. 4. Bevelled micropipette with a microforge-sharpened tip.

A cell larger in diameter than the mouth of the micropipette is selected for transfer so that the plasma membrane is ruptured as the cell enters the pipette. Too small a cell will not rupture, and upon transplantation, it will not participate in development. A cell too large for the micropipette will damage the donor nucleus so care must be taken to select cells and micropipettes of the appropriate size.

A recipient ovum (previously activated, enucleated, and dejellied) is placed in a depression in an operating dish filled with transplantation medium (McKinnell, 1978), and the micropipette, which contains the donor cell, is introduced into the

Fig. 5. Nuclear transplantation equipment consisting of a microinjection apparatus held in position with a micromanipulator.

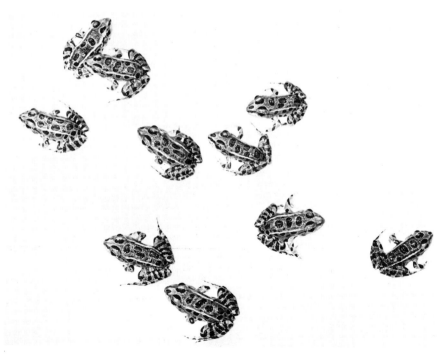

Fig. 6. Cloned juvenile *R. pipiens* produced by nuclear transplantation (photograph by Gordon A. F. Dunn).

ovum. If the micropipette tip is properly sharpened and if the experiment is conducted in a moderately cool laboratory (10 to 18°C), the surface of the recipient ovum will not be distorted excessively by the pressure of the micropipette tip. The broken donor cell is then deposited in the recipient ovum. Withdrawal of the micropipette is next. A channel is sometimes formed when the micropipette is removed. The channel, which permits leakage from the operated egg, can be severed with crossed microneedles (King, 1967). A successful nuclear transplanter is rewarded with one or more cloned frogs (Fig. 6). Methodology for amphibian cloning is described in adequate detail to permit experimentation in several publications (Elsdale *et al.*, 1960; King, 1966, 1967; Gurdon and Laskey, 1970; McKinnell, 1978).

III. CLONED FROGS

Significant biological information may result from nuclear transplantation experiments when less than a frog results subsequent to the transfer of a test nucleus to an enucleated ovum. Some of these experiments will be described in Section IV. However, less than a normal cloned adult would be totally useless for animal breeding purposes. How successful have nuclear transplantation experiments been in producing normal adult frogs?

Success, as defined by cloned frogs, depends upon the proper transfer of nuclei obtained from very young embryo donors. For example, a high proportion of recipient eggs that have received blastula nuclei develop normally (Briggs and King, 1952, 1953, 1955; Briggs, 1977). Nuclear transplant frogs are sometimes reared to sexual maturity as has been done for *R. pipiens* (Fig. 7) (McKinnell, 1962), *X. laevis* (Gurdon, 1962), as well as other species (Kawamura and Nishioka, 1977).

The yield of normal frogs in nuclear transfer operations diminishes rapidly as the age of the nuclear donor increases. The phenomenon of totipotency loss associated with nuclear age has been studied most in *R. pipiens* and *X. laevis* and probably is characteristic of all amphibian species (Gallien, 1966; King, 1979; McKinnell, 1972, 1978, 1979a; DiBerardino and Hoffner, 1980). The reason(s) for the loss of totipotency is unknown. Whether or not the phenomenon is reversible continues now, as in the past, to stimulate a number of cloning studies (see Sec. V).

Nuclear markers provide a high level of confidence that the normal development reported for cloned frogs is attributable to the inserted nucleus and is not due to gynogenetic development because of inadvertently retained maternal (egg) chromosomes. Examples of genetic markers that have proved useful or are potentially useful in cloning experiments include pigment mutants Burnsi (Simpson and McKinnell, 1964), Kandiyohi (McKinnell, 1960, 1964), albinism (Signoret *et al.*, 1962; Hoperskaya, 1975), and Pale (Ellinger and Carlson, 1978; Ellinger,

Fig. 7. Sexually mature male cloned frog with two of his many progeny (from McKinnell, 1962).

1980), polyploidy (Gallien *et al.*, 1963; McKinnell *et al.*, 1969a), and nucleolar mutant (Elsdale *et al.*, 1958). The reader in search of other potentially useful mutant genes should refer to one or more of the following reviews: Briggs (1973), Malacinski and Brothers (1974), Browder (1975), Gurdon and Woodland (1975), and Humphrey (1975).

Recipient ova are obtained that are known to be free of the genetic marker. A donor nucleus with the marker is inserted into the prepared ovum. Expression of the genetic marker witnesses to the success of nuclear transplantation, and together with the time of cleavage and chromosome counts, provides evidence of ovum enucleation.

Many species of amphibians have been studied by nuclear transfer (McKinnell, 1978). Recently fish have been cloned. Feeding larvae were obtained in the USSR when blastula nuclei were transplanted into ultraviolet enucleated ova of the loach, *Misgurnus fossilis* (Gasaryan *et al.*, 1979) and sexually mature hybrids from nuclei of carp and enucleated cytoplasm of the crucian were obtained in China (Research Group of Cytogenetics, Academy Sinica *et al.*, 1980). Mature nuclear transplant animals thus far have been reported in teleosts, amphibians, and mice. Recent results with mammalian cloning experiments are reported in Section VI.

IV. NUCLEI OF ADULT CELLS FAIL TO PROGRAM FOR NORMAL DEVELOPMENT

Thus far, adult cell nuclei have not been shown to be able to substitute for a zygote nuclei in development in contrast to nuclei obtained from very young embryos. The rationale for nuclear transplantation that provided the impetus for the original cloning studies persists in investigations of nuclei obtained from highly specialized cells. There are two alternative possibilities that may explain why adult cell nuclei cannot be cloned: First, the differentiated state of maturing and adult organisms is attributable to a genome held in common by all cell types of an organism, and the uniqueness of individual cell types is due to differential gene function. Associated with this view is the observation that the differentiated state is highly stable, and egg cytoplasm is an inadequate milieu to reprogram specialized nuclei. Second, highly specialized cells may owe their differentiated state to altered DNA structure (Robertson, 1980). The first alternative is sometimes referred to as the "theory of nuclear equivalence." It suggests that with proper procedures (not presently available), somatic nuclei from adult cells will be able to replace the zygote nucleus. The second alternative mandates that regardless of treatment, alterations in DNA structure accompanying cell specialization irrevocably preclude somatic nuclei from adult cells substituting for a zygote nucleus in development.

A number of studies have sought resolution of this fundamental question in developmental biology. Adult gut nuclei (McAvoy *et al.*, 1975), differentiated skin cell nuclei (Gurdon *et al.*, 1975), nuclei of immunoglobulin-producing lymphocytes (Wabl *et al.*, 1975; DuPasquier and Wabl, 1977), erythrocyte and erythroblast nuclei (Brun, 1978), and tumor-cell nuclei obtained from adult frogs (King and McKinnell, 1960; DiBerardino and King, 1965; King and DiBerardino, 1965; McKinnell *et al.*, 1969a, 1976; McKinnell, 1979b) share in common the capacity, when transplanted to enucleated egg cytoplasm, to evoke limited embryonic development. None of these experiments has yielded an adult frog. Even the stem cells of the adult male gamete fail to provide nuclei capable, with present techniques, of yielding cloned adult frogs. A transplanted nucleus of a spermatogonial cell produced a feeding larva that lived for 20 days and then died (DiBerardino and Hoffner, 1971). The generalization to be drawn from these studies is that the specialized state of adult nuclei, for whatever reason, precludes full participation in developmental processes after transplantation.

While the ultimate reason for failure of the inserted nucleus to function completely in development is not known, certainly the proximal cause of abnormal development and early death is known. The cell cycle of cleaving eggs is short compared to the cell cycle of specialized cells. The failure of the cell cycle of the inserted nucleus to become coordinated with the division cycle of the host egg results in chromosomal abnormalities that cause the nuclear transplant's demise (DiBerardino and Hoffner, 1970; DiBerardino, 1979). The relationship of cell cycle and transplantability of nuclei has also been studied by Ellinger (1978) and Beroldingen (1981).

There is a possible exception to the results of the experiments just described. Frogs that developed to metamorphosis were reported with nuclear transfer of cultured adult lens cells (Muggleton-Harris and Pezzella, 1972). It seems that because the lens nuclear transfers are an exception to the results obtained from other widely separated laboratories, the investigators ought to use one or more genetic markers associated with the donor nucleus to provide assurance that their results are valid.

V. PROCEDURES THAT ENHANCE DEVELOPMENTAL POTENTIALITY

If failure to demonstrate totipotency in transplanted nuclei is due to inadequate knowledge of the control of gene expression and is not due to permanently altered DNA structure, then there is at least the rational expectation that future studies will result in successful transfer of adult nuclei.

For successful adult cell nuclear transfer, it is necessary to reactivate many genes that have been silent for most of the life of the organism and to repress active genes that are associated with the highly specialized state of a differentiated cell. This activation and repression must be coordinated in a temporally appropriate manner for normal development. This is an extraordinarily difficult demand. Most likely, success will be due to some enhancement of the already extensive capability of egg cytoplasm to redirect gene activity.

Two examples will perhaps serve to illustrate the possibility of provoking the participation of nuclei that otherwise could not take part in development after transplantation. Endoderm nuclei of *R. pipiens* obtained from tailbud stage nuclear donors rarely would be expected to promote normal development with conventional nuclear transfer procedure. However, if spermine is added to the nuclear transfer medium and if the temperature of the operating room is lowered to about 10°C, 23% of tailbud endoderm nuclear transfers develop normally (Hennen, 1970). In this particular study, the specific action of spermine, an ubiquitous polycationic amine, is not understood. However, it is clear from the experiment that at least some restrictions of developmental potentiality are reversible.

It has been known for several years that oocyte cytoplasm has a profound effect on somatic nuclei. The oocyte's extraordinarily large nucleus, the germinal vesicle, undergoes no DNA synthesis. Rather, RNA synthesis occurs in the oocyte during its prolonged growth period. Somatic nuclei, when placed in growing oocyte cytoplasm, mimic germinal vesicles in several significant ways. The inserted nucleus increases enormously in volume, and there is an enhancement of RNA synthesis (Gurdon, 1967, 1968; Gurdon *et al.*, 1979). Would a somatic nucleus, otherwise not competent to promote embryonic development, be reprogrammed after a period of residence in oocyte cytoplasm? A partial answer to that question is provided by a study of late tailbud endoderm nuclei inserted into oocytes at first meiotic metaphase. The host oocyte, with its transplanted endoderm nucleus, was activated 24 hr later, and the maternal nucleus was removed surgically. About one-half of the operated oocytes formed blastulae, 21% gastrulated, 5% formed neurulae, and 1 (1%) developed into a larva (Hoffner and DiBerardino, 1980). The results of this experiment suggest, for the purposes of this paper, that the genome of a differentiating cell can respond to the powerful influence of oocyte cytoplasm, survive activation of the oocyte, and participate in embryonic development. Oocyte cytoplasm treatment could be used as part of a system to condition or reprogram somatic nuclei from differentiated cells.

VI. NUCLEAR TRANSPLANTATION IN MAMMALS

What are the biological requirements for nuclear transfer in mammals? Little more is required than that for cloning with amphibians: a source of eggs, a supply of somatic donor nuclei, an enucleation procedure, and a technique to put the two together. The operated egg must be cultured *in vitro* (Biggers *et al.*, 1971) until it is ready to be transferred to an appropriate host. The cultured cloned embryo then must be transferred to a hospitable uterus (Seidel, 1981). Because of these relatively simple requirements, it is not surprising that substantial progress has been made in the past few years, and mice produced by nuclear transplantation have been reported (Illmensee and Hoppe, 1981).

Ova are easily obtained by hormone treatment from most mammals including prepuberal (Gates, 1956) and mature mice (Fowler and Edwards, 1957; Allan and McLaren, 1971), shrews (Rutter and McKinnell, 1973), lions (Rowlands and Sadleir, 1968), squirrel monkeys (Bennett, 1967; Dukelow, 1970), cows (Elsden *et al.*, 1978), and humans (Edwards and Steptoe, 1975).

Asterias eggs, which are about the same size as many mammalian eggs, were enucleated many years ago by sucking out meiotic spindles (McClendon, 1908). Haploid mouse embryos were similarly produced by microsurgical removal of one pronucleus (Modlinski, 1975). Homozygous diploid embryos, produced by the extraction of one pronucleus and a delay in cleavage induced with cytochalasin B, can also be produced microsurgically (Hoppe and Illmensee, 1977; Markert and Petters, 1977). Removal of the maternal pronucleus from an unfertilized egg or both pronuclei from an inseminated ovum could produce an ovum suitable as a recipient for a transplanted nucleus. The latter procedure, i.e., the sucking out of both pronuclei of fertilized eggs, was used in successful mouse nuclear transplantation (Illmensee and Hoppe, 1981).

Cells from a donor embryo can be obtained by treating it with pronase to digest the enclosing zona pellucida (Mintz, 1962), or it can be removed mechanically (Illmensee and Hoppe, 1981). The exposed embryo can then be dissociated enzymatically. The dissociated cells now need only to be joined to the enucleated ovum.

There seem to be at least two means of inserting the donor nucleus into the recipient cytoplasm. One means relates to the fact that when cells of two types are cultured together, spontaneous fusion sometimes occurs (Barski *et al.*, 1961; Weiss, 1980). The Sendai virus, added to the culture, increases the rate of fusion many hundreds of times (Okada, 1958). The virulence of the virus is minimized by ultraviolet irradiation or chemical inactivation (Rao and Johnson, 1972; Watkins, 1973). Mouse spleen cells (Graham, 1971), fibroblasts (Baranska and Koprowski, 1970), lymph node cells, and bone marrow cells (Lin *et al.*, 1973) were fused to mouse eggs with the aid of the Sendai virus. Although these studies demonstrate the possibility of exploiting viruses to assist in mammalian cloning, in fact, the operated egg underwent only a few cleavage divisions.

Microsurgery is a more direct method for effecting a union of somatic nucleus with recipient ovum. The mouse egg (Lin, 1971), as well as the early embryo (Rossant *et al.*, 1978; Dewey and Mintz, 1980; Illmensee, 1980), obviously can survive much surgical manipulation. Pioneer microsurgical nuclear transplantation studies provided hope of eventual success with mammalian cloning (Bromhall, 1975; Modlinski, 1978). Success came with the report of three live-born mice produced by nuclear transplantation (Illmensee and Hoppe, 1981). While there seems to be little justification, if any, for the effort (McKinnell, 1979a), there is on record a human cloning attempt (Shettles, 1979).

It is not the purpose of this brief discourse to outline how to clone a mammal. Rather, the discourse is provided to indicate that procedures are available with mammals that are entirely analogous to the methods that have resulted in successful cloning with amphibians. Theoretically, routine cloning, at least of embryonic mammalian cell nuclei, can be reasonably anticipated in the future. Whether or not cloning of adult nuclei becomes possible depends upon increased understanding of the differentiative process.

ACKNOWLEDGMENTS

I thank Marie A. DiBerardino and Robert D. Bergad for helpful comments in the preparation of the manuscript. The experimental work of the author has been supported by grants from the National Institutes of Health and the National Science Foundation.

REFERENCES

Allen, J., and McLaren, A. (1971). *J. Reprod. Fertil.* **27**, 137–140.
Baranska, W., and Koprowski, H. (1970). *J. Exp. Zool.* **174**, 1–14.
Barski, G., Sorievl, S., and Cornefert, F. (1961). *J. Natl. Cancer Inst.* **26**, 1269–1291.

Battin, W. T. (1959). *Exp. Cell Res.* **17**, 59-75.

Bennett, J. P. (1967). *J. Reprod. Fertil.* **13**, 357-359.

Biggers, J. D., Whitten, W. K., and Whittingham, D. G. (1971). *Methods Mamm. Embryol.*, pp. 86-116.

Briggs, R. (1973). *In* "Genetic Mechanisms in Development" (F. H. Ruddle, ed.), pp. 169-199. Academic Press, New York.

Briggs, R. (1977). *In* "Cell Interactions in Differentiation" (M. Karkinen-Jääskeläinen, L. Saxén, and L. Weiss, eds.), pp. 23-43. Academic Press, New York.

Briggs, R., and King, T. J. (1952). *Proc. Natl. Acad. Sci. U.S.A.* **38**, 455-463.

Briggs, R., and King, T. J. (1953). *J. Exp. Zool.* **122**, 485-505.

Briggs, R., and King, T. J. (1955). *In* "Biological Specificity and Growth" (E. G. Butler, ed.), pp. 207-228. Princeton Univ. Press, Princeton, New Jersey.

Briggs, R., Signoret, J., and Humphrey, R. R. (1964). *Dev. Biol.* **10**, 233-246.

Bromhall, J. D. (1975). *Nature (London)* **258**, 719-722.

Browder, L. W. (1975). *Handb. Genet.* **4**, 19-33.

Brun, R. B. (1978). *Dev. Biol.* **65**, 271-284.

Dewey, M. J., and Mintz, B. (1980). *In* "Differentiation and Neoplasia" (R. G. McKinnell, M. A. DiBerardino, M. Blumenfeld, and R. Bergad, eds.), pp. 275-282. Springer-Verlag, Berlin and New York.

DiBerardino, M. A. (1979). *Int. Rev. Cytol. Suppl.* **9**, 129-160.

DiBerardino, M. A. (1980). *Differentiation (Berlin)* **17**, 17-30.

DiBerardino, M. A., and Hoffner, N. (1970). *Dev. Biol.* **23**, 185-209.

DiBerardino, M. A., and Hoffner, N. (1971). *J. Exp. Zool.* **176**, 61-72.

DiBerardino, M. A., and Hoffner, N. J. (1980). *In* "Differentiation and Neoplasia" (R. G. McKinnell, M. A. DiBerardino, M. Blumenfeld, and R. Bergad, eds.), pp. 53-74. Springer-Verlag, Berlin and New York.

DiBerardino, M. A., and King, T.J. (1965). *Dev. Biol.* **11**, 217-242.

Dukelow, W. R. (1970). *J. Reprod. Fertil.* **22**, 303-309.

DuPasquier, L., and Wabl, M. R. (1977). *Differentiation (Berlin)* **8**, 9-19.

Edwards, R. G., and Steptoe, P. C. (1975). *In* "Progress in Infertility" (S. J. Behrman and R. W. Kistner, eds.), 2nd ed., pp. 377-409. Little, Brown, Boston.

Ellinger, M. S. (1978). *Dev. Biol.* **65**, 81-87.

Ellinger, M. S. (1979). *J. Morphol.* **162**, 77-92.

Ellinger, M. S. (1980). *J. Embryol. Exp. Morphol.* **56**, 125-137.

Ellinger, M. S., and Carlson, J. T. (1978). *J. Exp. Zool.* **205**, 353-360.

Ellinger, M. S., King, D. R., and McKinnell, R. G. (1975). *Radiat. Res.* **62**, 117-122.

Elsdale, T. R., Fischberg, M., and Smith, S. (1958). *Exp. Cell Res.* **14**, 642-643.

Elsdale, T. R., Gurdon, J. B., and Fischberg, M. (1960). *J. Embryol. Exp. Morphol.* **8**, 437-444.

Elsden, R. P., Nelson, L. D., and Seidel, G. E., Jr. (1978). *Theriogenology* **9**, 17-26.

Fowler, R. E., and Edwards, R. G. (1957). *J. Endocrinol.* **15**, 374-384.

Gallien, L. (1966). *Ann. Biol.* **5-6**, 241-269.

Gallien, L., Picheral, B., and Lacroix, J.-C. (1963). *C. R. Hebd. Seances Acad. Sci.* **256**, 2232-2234.

Gasaryan, K. G., Hung, N. M., Neyfakh, A. A., and Ivanenkov, V. V. (1979). *Nature (London)* **280**, 585-587.

Gates, A. H. (1956). *Nature (London)* **177**, 754-755.

Graham, C. F. (1971). *Acta Endocrinol. Suppl.* **153**, 154-167.

Gurdon, J. B. (1960). *Q. J. Microsc. Sci.* **101**, 299-311.

Gurdon, J. B. (1962). *Dev. Biol.* **4**, 256-273.

Gurdon, J. B. (1967). *In* "Methods in Developmental Biology" (F. H. Wilt and N. K. Wessells, eds.), pp. 75-84. Crowell-Collier, New York.

Gurdon, J. B. (1968). *Essays Biochem.* **4**, 25-68.

Gurdon, J. B., and Laskey, R. A. (1970). *J. Embryol. Exp. Morphol.* **24**, 249-255.

Gurdon, J. B., and Woodland, H. R. (1975). *Handb. Genet.* **4**, 35-50.

Gurdon, J. B., Laskey, R. A., and Reeves, D. R. (1975). *J. Embryol. Exp. Morphol.* **34**, 93-112.

Gurdon, J. B., Laskey, R. A., DeRobertis, E. M., and Partington, G. A. (1979). *Int. Rev. Cytol. Suppl.* **9**, 161-178.

Hennen, S. (1970). *Proc. Natl. Acad. Sci. U.S.A.* **66**, 630-637.

Hennen, S. (1973). *J. Embryol. Exp. Morphol.* **29**, 529-538.

Hoffner, N. J., and DiBerardino, M. A. (1980). *Science (Washington, D.C.)* **209**, 517-519.

Hoperskaya, O. A. (1975). *J. Embryol. Exp. Morphol.* **34**, 253-264.

Hoppe, P. C., and Illmensee, K. (1977). *Proc. Natl. Acad. Sci. U.S.A.* **74**, 5657-5661.

Humphrey, R. R. (1975). *Handb. Genet.* **4**, 3-17.

Illmensee, K. (1980). *In* "Differentiation and Neoplasia" (R. G. McKinnell, M. A. DiBerardino, M. Blumenfeld, and R. Bergad, eds.), pp. 75-92. Springer-Verlag, Berlin and New York.

Illmensee, K., and Hoppe, P. C. (1981). *Cell* **23**, 9-18.

Jones, K. W., and Elsdale, T. R. (1963). *J. Embryol. Exp. Morphol.* **11**, 135-154.

Kawamura, T., and Nishioka, M. (1963). *J. Sci. Hiroshima Univ., Ser. B, Div. 1* **21**, 1-13.

Kawamura, T., and Nishioka, M. (1977). *Sci. Rep. Lab. Amphib. Biol., Hiroshima Univ.* **2**, 1-23.

King, T. J. (1966). *Methods Cell Physiol.* **2**, 1-36.

King, T. J. (1967). *In* "Methods in Developmental Biology" (F. H. Wilt and N. K. Wessells, eds.), pp. 737-751. Crowell-Collier, New York.

King, T. J. (1979). *Int. Rev. Cytol. Suppl.* **9**, 101-106.

King, T. J., and DiBerardino, M. A. (1965). *Ann. N. Y. Acad. Sci.* **126**, 115-126.

King, T. J., and McKinnell, R. G. (1960). *In* "Cell Physiology of Neoplasia," pp. 591-617. Univ. of Texas Press, Austin.

Lin, T. P. (1971). *In* "Methods in Mammalian Reproduction"(J. C. Daniel, Jr., ed.), Academic ·

Lin, T. P., Florence, J., and Oh, J. O. (1973). *Nature (London)* **242**, 47-49.

McAvoy, J. W., Dixon, K. E., and Marshall, J. A. (1975). *Dev. Biol.* **45**, 330-339.

McClendon, J. F. (1908). *Wilhelm Roux' Arch. Entwicklungsmech. Org.* **26**, 662-668.

McKinnell, R. G. (1960). *Am. Nat.* **94**, 187-188.

McKinnell, R. G. (1962). *J. Hered.* **53**, 199-207.

McKinnell, R. G. (1964). *Genetics* **49**, 895-903.

McKinnell, R. G. (1972). *In* "Cell Differentiation" (R. Harris, P. Allin, and D. Viza, eds.), pp. 61-64. Munksgaard, Copenhagen.

McKinnell, R. G. (1978). "Cloning, Nuclear Transplantation in Amphibia." Univ. of Minnesota Press, Minneapolis.

McKinnell, R. G. (1979a). "Cloning, A Biologist Reports." Univ. of Minnesota Press, Minneapolis.

McKinnell, R. G. (1979b). *Int. Rev. Cytol. Suppl.* **9**, 179-188.

McKinnell, R. G., and Schultheis, M. (1979). *Am. Assoc. Lab. Anim. Sci. Pub.* **79-4** (Abstr. P11).

McKinnell, R. G., Deggins, B. A., and Labat, D. D. (1969a). *Science (Washington, D.C.)* **165**, 394-396.

McKinnell, R. G., Mims, M. F. and Reed, L. A. (1969b). *Z. Zellforsch. Mikrosk. Anat.* **93**, 30-35.

McKinnell, R. G., Steven, L. M., Jr., and Labat, D. D. (1976). *In* "Progress in Differentiation Research" (N. Müller-Bérat, ed.), pp. 319-330. North-Holland Publ., Amsterdam.

McKinnell, R. G., Gorham, E., Martin, F. B., and Schaad, J. W. (1979). *Lab. Anim. Sci.* **29**, 66-68.

McKinnell, R. G., Kren, B. T., Bergad, R., Schultheis, M., Byrne, T., and Schaad, J. W. (1981). *Teratogenesis, Carcinogenesis, and Mutagenesis* (in press).

Malacinski, G. M., and Brothers, A. J. (1974). *Science (Washington, D.C)* **184**, 1142–1147.

Markert, C. L., and Petters, R. M. (1977). *J. Exp. Zool.* **201**, 295–302.

Mintz, B. (1962). *Science (Washington, D.C.)* **138**, 594–595.

Modlinski, J. A. (1975). *J. Embryol. Exp. Morphol.* **33**, 897–905.

Modlinski, J. A. (1978). *Nature (London)* **273**, 466–467.

Moscona, A. (1952). *Exp. Cell Res.* **3**, 535–539.

Muggleton-Harris, A. L., and Pezzella, K. (1972). *Exp. Gerontol.* **7**, 427–431.

Okada, Y. (1958). *Biken J.* **1**, 103–110.

Porter, K. R. (1939). *Biol. Bull.* **77**, 233–257.

Rao, P. N., and Johnson, R. T. (1972). *Methods Cell Physiol.* **5**, 75–126.

Research Group of Cytogenetics, Institute of Zoology, Academy Sinica; Research Group of Somatic Cell Genetics, Institute of Hydrobiology, Academy Sinica; and Research Group of Nuclear Transplantation, Chang Jiang Fisheries Research Institute, State Fisheries General Board (1980). *Sci. Sin.* (Eng. Ed.) **23**, 517–525.

Robertson, M. (1980). *Nature (London)* **287**, 390–392.

Rossant, J., Gardner, R. L., and Alexandre, H. L. (1978). *J. Embryol. Exp. Morphol.* **48**, 239–247.

Rowlands, I. W., and Sadleir, R. M. (1968). *J. Reprod. Fertil.* **16**, 105–111.

Rugh, R. (1934). *Biol. Bull.* **66**, 22–29.

Rutter, P. A., and McKinnell, R. G. (1973). *Am. Zool.* **13**, 1312–1313.

Sambuichi, H. (1959). *J. Sci. Hiroshima Univ., Ser. B, Div. 1* **18**, 39–43.

Seidel, G. E., Jr. (1981). *Science (Washington, D.C.)* **211**, 351–358.

Shettles, L. B. (1979). *Am. J. Obstet. Gynecol.* **133**, 222–225.

Signoret, J., Briggs, R., and Humphrey, R. R. (1962). *Dev. Biol.* **4**, 134–164.

Simpson, N. S., and McKinnell, R. G. (1964). *J. Cell Biol.* **23**, 371–375.

Subtelny, S., and Bradt, C. (1960). *Dev. Biol.* **2**, 393–407.

Subtelny, S., and Bradt, C. (1963). *J. Morphol.* **112**, 45–59.

Townes, P. L., and Holtfreter, J. (1955). *J. Exp. Zool.* **128**, 53–120.

von Beroldingen, C. H. (1981). *Dev. Biol.* **81**, 115–126.

Wabl, M. R., Brun, R. B., and DuPasquier, D. (1975). *Science (Washington, D.C.)* **190**, 1310–1312.

Watkins, J. F. (1973). *In* "Seventh National Cancer Conference Proceedings," pp. 61–63. Lippincott, Philadelphia.

Weiss, M. C. (1980). *In* "Differentiation and Neoplasia" (R. G. McKinnell, M. A. DiBerardino, M. Blumenfeld, and R. Bergad, eds.), pp. 87–92. Springer-Verlag, Berlin and New York.

Wright, P. A., and Flathers, A. R. (1961). *Proc. Soc. Exp. Biol. Med.* **106**, 346–347.

10

Parthenogenesis, Identical Twins, and Cloning in Mammals

CLEMENT L. MARKERT AND GEORGE E. SEIDEL, JR.

I. INTRODUCTION

Expanding knowledge in molecular genetics now indicates that DNA and RNA are far more labile than was supposed a decade ago (Abelson, 1980).

181

NEW TECHNOLOGIES IN ANIMAL BREEDING
Copyright © 1981 by Academic Press, Inc.

Moreover, our increasing ability to modify and manipulate the genome has blurred the distinction between sexual and asexual classifications of reproduction. Nevertheless, sexual reproduction, as classically defined, has nearly always been the winner in evolution (Maynard Smith, 1978). All organisms compete, and most species that did not combat sex with sex became extinct, primarily because of being crowded out by the winners (Maynard Smith, 1978). Sexual reproduction is obviously important in achieving a rapid adaptation to climatic and other physical changes in the environment as well as in achieving competitive fitness.

In the face of this knowledge, the present great interest in methods of reproducing mammals asexually is perhaps ironic. In any case, there is a lively interest in being able to produce exact genetic copies of outstanding domestic animals, and an exaggerated fear of the same prospect for human beings. Preliminary analyses suggest that asexual reproduction would not be as great a boon for animal breeders as one might think (Seidel, 1980; and Chapter 12, this volume). Nevertheless, there would be some advantages. The value of such animals for research would be tremendous; this has been proven repeatedly in laboratory rodents, resulting in the use of millions of inbred animals and their F_1 crosses each year. Many kinds of experiments would be impossible without genetically defined stocks, even though the animals may suffer from reduced vitality due to homozygosity (except for F_1 crosses).

The relationships among parthenogenesis, identical twins, and cloning may not be obvious at first. However, in nonhomozygous animals, identical twins (or triplets, etc.) are the ultimate clone in the sense of genetic identity. Further, identical twins can already be produced experimentally with reasonable reliability. Parthenogenesis, a method of selfing, produces offspring fairly similar to their mothers, but by no means genetically identical. An important point with parthenogenesis is that in most species in which it occurs, the oocyte is activated without participation of a sperm. This problem of activation must be dealt with in some methods of cloning. The function of sperm in activation of the oocyte is not at all well understood in mammals, and the study of parthenogenesis may provide information crucial to successful cloning.

Another interesting and perplexing problem with mammalian parthenotes is that they fail to develop to term, even when diploid and derived from highly inbred lines, which therefore do not have recessive deleterious alleles. The observation that cells of parthenogenetic origin form normal adult tissue when chimeras are made between parthenogenetic and normal embryos makes this problem even more interesting (Stevens et al., 1977; Surani et al., 1977; Stevens, 1978). If some as yet unknown biological barrier blocks successful parthenogenetic development, then such barriers will surely have profound consequences for cloning mammals.

II. PARTHENOGENESIS

A. Occurrence in Nature

Parthenogenesis is defined as reproduction without the genetic participation of the sperm. It is a very common method of reproduction, present in most phyla of animals. It occurs in all vertebrate classes except mammals, although rarely in birds. Animal parthenogenesis has been reviewed by Uzzell (1970), Cuellar (1977), and Gerritsen (1980).

The most studied avian models of parthenogenesis are certain strains of turkeys. Several mechanisms have been suggested over the years, but the most recent evidence indicates that development begins with haploid cells (DeFord et al., 1979). Somehow, diploid cells arise, probably by lack of cytokinesis after DNA replication, and the progeny of these cells (or cell) give rise to the organism proper. Such organisms would have to be completely homozygous and hence male (in birds), which is what is found. Thus, one cannot dispense with the two sexes with this mode of parthenogenesis because the females that produce eggs must be obtained sexually. The parthenogenetic males are fertile.

If immediate cleavage of the secondary oocyte instead of extrusion of the second polar body were the parthenogenetic mechanism, chimeric males would be derived from the haploid cells. The separation of the Z and W sex chromosomes as bivalents at the first meiotic division precludes the formation of male–female sex chimeras. In birds, females are the heterogametic sex and all embryos with a WW sex chromosome constitution die early in development.

Parthenogenesis occurs naturally in some populations of reptiles and amphibia, in which the female lays eggs that develop into other females that lay eggs to make still more females without any male involvement (Beatty, 1967; Uzzell, 1970; Cuellar, 1971). Parthenogenesis occurs in many species of fish and in various ways: some produce ova that develop without male involvement; other species require a male of a related species to provide sperm to activate ovum development, but the sperm makes no genetic contribution; such fish are a kind of sexual parasite (Beatty, 1967). In any case, parthenogenesis is a common phenomenon, particularly among invertebrates, but also among vertebrates, although it has never been known to result in young mammals.

Many steps on the way to the formation of a normal haploid gamete can be altered to lead to parthenogenesis. In invertebrates, meiosis may be suppressed altogether so that the potential gamete, a diploid oogonium, can just continue to develop into an individual (Suomalainen, 1950). Alterations later in the process of gametogenesis can cause the second polar body either to re-fuse with the ovum to fertilize it, or not to be extruded in the first place. Either scheme results in diploid parthenotes. Since these and many other perturbations of meiosis occur in nature, they are not biologically impossible. If enough were known about meiosis, parthenogenetic mammals could probably be produced too. But, there is a long way to go.

Meiosis in the mouse oocyte begins with the breakdown of the germinal vesicle. This takes only a few minutes, once it starts (Sorensen, 1973; Kaplan *et al.*, 1978). Next, chromosomes begin to appear and line up on the spindle. At this point, it is possible to aspirate chromosomes with a micropipette, and it is also possible to put them back in, foreshadowing some distant day when we can rearrange the chromosome makeup of animals by surgery. Next, the first polar body is extruded. An ovum at this stage has a diploid DNA chromosome complement. If the ovum cleaved into two blastomeres immediately, instead of the second polar body being extruded, then haploid parthenogenetic development would ensue. Such parthenotes would be chimeric because the two blastomeres would not be identical owing to crossing over. Diploidy might be restored and chimerism eliminated by hydrostatic pressure, which in amphibians suppresses the second polar body or the first cleavage division. Perhaps cytochalasin B might also be used to achieve the same end (Tsukahara and Ishikawa, 1980). Clearly, further research is needed in this important area (Tompkins, 1978; Gillespie and Armstrong, 1979; Reinschmidt *et al.*, 1979).

To achieve parthenogenesis, particularly by the complete suppression of meiosis, would require an understanding of the basic events in meiosis, including mechanisms of chromosome pairing. Suppressing meiosis and activating the oocyte to develop without involving sperm would be the ideal way to produce clones of animals. Doing this in a cow, for example, could lead to the production of very valuable dairy herds without the need for any bulls and without the uncertainties of sexual reproduction.

B. Activation of Oocytes

At fertilization, oocytes of most vertebrates are released from a state of metabolic and developmental quiescence by the process of activation. The second polar body is extruded at this time, and embryonic development begins. This may occur even in the absence of sperm, but the fertilizing sperm is usually the activating agent. A reasonable physiological approach to inducing parthenogenesis is to activate oocytes with sperm that have been damaged so that they cannot make a genetic contribution. In amphibians, this can be done easily by simply irradiating the sperm with ultraviolet light or X rays (Hertwig, 1911; Rugh, 1939; Tompkins, 1978). One can obtain literally thousands of gynogenetic haploid embryos with this method because the irradiated sperm provoke frog ova to develop but do not make any genetic contribution. One would imagine that this could also be done with mammals, particularly with *in vitro* fertilization. However, success to date has been middling at best. It is possible to treat mammalian sperm with ultraviolet light or chemicals like mitomycin C to inactivate the genetic material, but the results are not like those with amphibia. Mouse eggs fertilized with irradiated sperm are activated but early development is abnormal (Edwards, 1957; Bedford and Overstreet, 1972; Thadani, 1981). When mammalian sperm are irradiated heavily, they seem incapable of fertilizing the oocyte. And if the sperm are not irradiated sufficiently, the sperm contribute abnormal genetic material, which leads to abnormal embryonic development. To date, no one has succeeded in treating mammalian sperm so that they initiate parthenogenesis. Theoretically, it still seems that this approach should succeed, but perhaps species differences are so great that this will not be possible with mammals.

There are numerous other ways of provoking mammalian ova to develop parthenogenetically including the use of various chemicals, calcium ionophore (AY23187), electric shock, ethanol, and abrupt temperature change (Graham, 1974; Steinhart et al., 1974). Recently Soupart et al. (1978) have initiated parthenogenetic development by fusing mouse oocytes with Sendai virus, in effect fertilizing one ovum with another. With some strains of mice spontaneous parthenogenesis is fairly common without any obvious external stimulus (Eppig et al., 1977). These mice just produce ova, a large fraction of which start to develop without being fertilized. Most of the parthenotes produced as a consequence of artificial activation, e.g., electrical shock to the oviduct, start out as haploids and, therefore, would not be expected to develop very far. Many of those that occur naturally in mice, however, are diploids, presumably because of the suppression of the second polar body (Kaufman et al., 1977). Some of these will develop extensively, a few to the 25-somite stage. However, none has survived to birth. From the moment the parthenotes start developing, the fraction surviving diminishes, and finally the last one dies. This is surprising. There seems to be no biological reason why they should not survive; a 25-somite embryo is a very extensively developed individual made up of many kinds of highly specialized cells. It is hard to imagine what kind of defect from the initial parthenogenetic activation could be retained all the way to this very advanced stage. Yet, there is some problem because hundreds of such parthenogenetic mouse ova have been given the opportunity to develop and have failed to do so (Kaufman and Gardner, 1974). That the mammal is the only class in which parthenogenesis has never succeeded suggests that some barrier has been invented during evolution to make such development highly improbable. One possible explanation is that the sperm, in addition to activating the ovum and contributing the paternal genome, may make a third contribution essential to normal development. One approach that has been used to test this question is to take a normally fertilized ovum and remove the sperm microsurgically just after "fertilization." Although such haploid ova were activated naturally, they did not develop much better than other haploids. However, as will be described later, such embryos provide starting material for one method of cloning mammals.

III. IDENTICAL TWINS

A. Occurrence in Nature

Identical twins, triplets, etc. are natural examples of cloning from which much can be learned. The incidence of identical twin births in man and cattle is a few per thousand births; the frequency is influenced by race and breed and increases with maternal age (Johansson *et al.*, 1974). Identical triplets and quadruplets are probably exceedingly rare among most mammals, although the incidence of identical multiplets is unknown in most species because they go unnoticed due to phenotypic similarities among nonidentical members of a litter. Marmosets almost always have fraternal twins. The nine-banded armadillo usually has identical quadruplets and the eleven-banded armadillo has up to 10 or more identical young in each litter (Patterson, 1913). The mechanisms that induce naturally occurring identical twins, triplets, etc. remain relatively unstudied, although valuable clues have been obtained by examining placentas. For example, the majority of human identical twins have a common allanto-chorion but separate amnions.

A recent paper by Hsu and Gonda (1980) describes one mechanism of identical twin formation *in vitro*. When attachment of mouse embryos to plastic occurred via mural trophoblast directly opposite the inner cell mass, two egg cylinders formed because of mechanical constraints. One other recent paper (Kaufmann and O'Shea, 1978) describes conjoint (Siamese) monozygotic twin mice produced by vincristine injections during gestation. It is not yet clear whether either of these studies is related to mechanisms that generate naturally occurring identical twins or whether either of the experimental mechanisms can be exploited to increase the rate of identical twinning.

B. Experimental Production of Identical Twins

Clues that it might be possible to produce identical twins came from studies of totipotency. For example, Seidel (1952) destroyed one cell of 2-cell rabbit embryos and found that perfectly normal young resulted when these embryos were transferred to recipients. Extensions of this work have demonstrated that sometimes normal young result when three of four blastomeres are destroyed and even rarely when seven of eight are destroyed (Tarkowski and Wroblewska, 1967). Apparently, the first investigators actually to produce identical twins were

Mullen *et al.* (1970), who removed the zona pellucida from 2-cell mouse embryos and cultured each blastomere separately before transferring the resulting embryos back to recipients. Much higher success rates were achieved by Moustafa and Hahn (1978), who simply bisected morulae of mice microsurgically and then transferred the halves.

Recently, Willadsen and co-workers (1979, 1980, 1981) have separated blastomeres at the 2-cell to morulae stage of development and placed them into surrogate zonae pellucidae of agar or patched with agar. These zonae with their one-half or one-fourth embryos were then transferred to the oviduct for a few days, recovered, and the agar trimmed off. By this time the embryos can survive without an intact zona pellucida and may then be transferred to the uterus of a recipient for gestation to term. A number of identical multiplets have been produced this way in cattle, sheep, and horses.

Techniques for producing identical twins will have several kinds of application. The first that comes to the mind of livestock producers is obtaining twice as many offspring per embryo. A variant of this is to transfer one-half (or one-fourth) and freeze the other(s) to await phenotypic information like sex, growth rate, or milk production. When this information is obtained, a very rational decision can be made whether to discard or transfer the remainder of the frozen embryo. An interesting twist, of great value for certain experiments, would be to transfer a frozen identical twin to her twin sister that had matured. The ability to obtain inexpensive identical twins at will would greatly decrease costs for many experiments with animals because genetic variation would be eliminated, and hence the number of animals required per group would be reduced without sacrificing experimental power. A further experimental application concerns basic questions about how embryos develop with only one-half as many or even fewer cells than in the normal embryo (Tarkowski and Wroblewska, 1967; Fernández and Izquierdo, 1980).

IV. CLONING VIA HOMOZYGOSITY

A. Evolution of the Homozygous Diploid Idea

About the time that the experiments were being completed on microsurgical removal of sperm from ova after activation, Snow (1973) published a paper in which he treated 2-cell embryos with cytochalasin B overnight. This prevents cell division but does not prevent DNA synthesis and nuclear replication; thus, after treatment, two nuclei are found per cell. After one cell cycle, Snow washed out the cytochalasin B, and cell division and development then continued, resulting in tetraploid embryos. It did not take much thinking to realize that combining the cytochalasin B treatment with microsurgical removal of sperm would diploidize the haploids (Markert and Petters, 1977). Moreover, such embryos would be perfectly homozygous. The resulting two haploid nuclei, identical to one another, would contribute their chromosomes to a single spindle when the nuclear membranes broke down at the next cell cycle. Thus, the diploid number of chromosomes per nucleus would be restored to produce a normal, but homozygous individual. Only females could be produced because homozygous YY genotypes are lethal; furthermore, if a single female can be defined as a strain

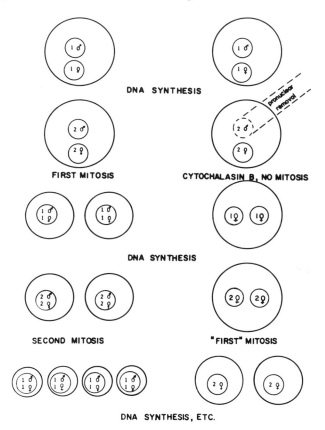

Fig. 1.   Steps in producing homozygous diploid embryos. The left-half of the figure depicts the normal series of events in early embryonic development. The larger circles represent cells; the small circles within them represent nuclei. Within the small circles are recorded the number of complements of maternal or paternal DNA. The right-half depicts the series of steps for making a homozygous diploid embryo. One of the pronuclei (in this example the one derived from the male) is removed microsurgically prior to the expected first mitosis. Cell division (but not nuclear division) is prevented by cytochalasin B treatment. Cytochalasin B is then removed, and the embryo undergoes the first cell division when normal embryos are undergoing the second cell division. This leads to restoration of the normal diploid condition, but with two female complements of DNA rather than one from each parent.

of mice, then strains of mice can be made in 3 weeks instead of the several years required by regimens of brother–sister mating. Females produced in this manner would not be genetically identical with their mother, because even highly inbred strains are not perfectly homozygous. Mutations occur in each strain, and some steady state level of heterozygosity must be established. This means that each ovum is slightly different from others from the same ovary. However, all the ova from a homozygous diploid female would by necessity be identical. Therefore, if the procedure were repeated with such ova, all of the resultant individuals would be genetically identical to one another and to the mother and would constitute a mammalian clone. These procedures are summarized in Fig. 1. The potential of such procedures with species that have long generation intervals is enormous. With cattle, for example, one could never hope by brother–sister mating to make an inbred strain in a human lifetime. With this microsurgical method, it could be done in one 9-month gestation. That might be a great boon for animal husbandry.

B. Mechanics of Producing Homozygous Diploids

The surgical techniques for making these homozygotes are not dissimilar to those for cloning by nuclear transplantation. Two micromanipulators are used to hold and move pipettes in the petri dish in which are placed fertilized eggs, one to each drop of medium under oil to prevent evaporation. The medium is equilibrated with a 5% CO_2 in air atmosphere. One of the pipettes is used to hold the embryo by suction on the zona pellucida, and the other one is for enucleation (See Figs. 1 and 2, in Chapter 6, this volume). It is very important to use a small enucleation pipette, with a very sharp bevel. With some practice, it is relatively easy to remove the pronuclei. Differential interference or phase contrast optics are desirable but not required for viewing pronuclei of common laboratory mammals. Pronuclei of embryos from cows and pigs are much more difficult to see. Pig oocytes have a very yolky cytoplasm that obscures the pronuclei. Nevertheless, even these ova can be enucleated by the same procedures that are successful with mice.

The ova to be enucleated are harvested some hours after fertilization when the two pronuclei become visible. The cumulus cells and corona radiata are removed with hyaluronidase treatment. The pronuclear membrane is a very tough, elastic, almost rubber-like membrane, which can be stretched and pulled completely through the plasma membrane while attached to the enucleation pipette by suction. The holding pipette is about 5–10 μm in diameter. One can aspirate the pronucleus directly into the enucleation pipette or just attach the pipette to the pronuclear membrane and pull it through the plasma membrane (Fig. 2). Removal of the whole pronucleus may be the better strategy, but this has not been studied critically. After enucleation, cytochalasin B at a dosage of 5 μg/ml is added to the medium to prevent cytokinesis; after culture overnight, embryos are

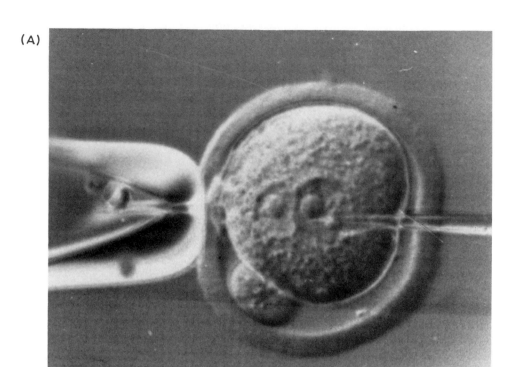

(A)

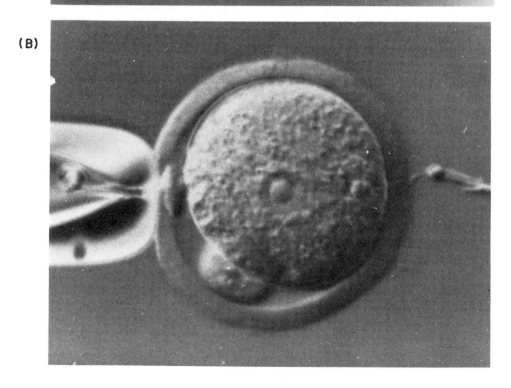

(B)

placed in medium without cytochalasin B. In practice, cytochalasin B is actually added one-half an hour prior to enucleation because this leads to higher survival rates of embryos after enucleation, probably because less damage results since fewer microfilaments are torn (Brown and Spudich, 1979). After the cytochalasin B is removed, the two pronuclei, now identical, break down and the chromosomes assemble on the spindle in preparation for the first cell division, which results in two normal diploid cells, one cell cycle behind normal (Fig. 1). Some of these develop into morulae and blastocysts *in vitro* (Market and Petters, 1977). For further development, they must be transferred to the reproductive tracts of recipient mice.

The experiment is set up genetically so that mistakes are easily identified. A homozygous black, non-agouti animal (aaCC) is crossed with an albino animal that carries homozygous alleles for agouti (AAcc) so that the offspring between the two genotypes will always be agouti in appearance. If the black pronucleus is removed, leaving the haploid albino nucleus behind, an albino animal would result. Conversely, if the albino nucleus is removed and the black nucleus left behind to be diploidized, a black animal will result. So one cannot be confused by the results. If enucleation fails, an agouti animal results, and if enucleation is successful, either an albino or black mouse results. Although one usually has great confidence in the actual enucleation, the genetic arrangement makes the results conclusive. Thousands of embryos have been enucleated in this way, but not many have developed to term (Hoppe and Illmensee, 1977; Modlinski, 1980; G. E. Seidel, Jr., and C. L. Markert, unpublished). Only Hoppe and Illmensee (1977) have reported the birth of such homozygous mice with "parents" of the genetic constitution just described. Two of the seven that they reported were derived from pronuclei contributed by the sperm—half of the nuclei contributed by sperm will carry the X chromosome and apparently can be diploidized to give rise to females from only the male parent. The remaining five offspring were from female pronuclei, showing that either pronucleus can be diploidized to make a completely normal individual. These mice have grown to adulthood and have reproduced.

C. Potential Theoretical Limitations

Why has cloning not been achieved by repeating the enucleation procedures on ova from homozygous females? Probably because the percentage of success is exceedingly low. Although thousands of homozygous embryos have been produced, these seven mice are the only ones so far reported in the literature to have been born. Clearly, there are problems with the viability of homozygous embryos

---

Fig. 2. Microsurgical removal of pronucleus from a mouse zygote. (A.) Zygote is held and micropipette introduced. (B.) Enucleation of one pronucleus completed. One nucleolus remained in the ooplasm. Such nucleoli soon disappear.

produced by the microsurgical techniques. However, one can in theory make clones of mammals this way. If homozygous cattle were produced, they could be cloned to yield absolutely identical offspring. Outstanding valuable clones could be perpetuated indefinitely.

One potential problem with some mammals is recessive lethal genes. Of course, this problem is circumvented in experiments with mice by using inbred strains, which are virtually homozygous already. Although information is scant, it does not appear that there are excessive frequencies of genetic lethals in most populations of domestic animals. To the extent that they are present, it is merely a statistical problem to obtain a haploid set of chromosomes without a lethal gene. For example, if a particular animal carries three recessive nonlinked lethals, then one in eight gametes would be free of these alleles.

Inbreeding depression, however, will be a serious problem. The mechanism of inbreeding depression is far from clear (Ginzburg, 1979). However, it does seem that every time two undesirable recessive alleles are paired, there should be an equal chance that two desirable ones pair, so that for every depression there ought to be an equivalent elevation. However, this seems not to be what happens. In terms of molecular biology, significant advantages of heterozygosity per se are not obvious. Optimal alleles should exist at each locus, and homozygosity of these alleles should, if anything, lead to genetically superior individuals. Some argue, to the contrary, that the heterozygote per se is better than either homozygote (Mukai *et al.,* 1964). Perhaps the heterozygosity provides a biological advantage during embryonic development by enabling different types of specialized cells to exploit one or the other product of the two alleles. However, there are no clear examples of genetic advantage to individuals possessing two allelic isozymes coded by allelic genes. One or the other zygote is superior or the heterozygotes are equivalent to the homozygotes. If heterozygote advantages do exist, they probably stem from differences in the regulatory DNA rather than in the structural genes themselves. Genetic control of the quantity of gene activity can be expressed over a wide range, certainly more than 100-fold for many enzymes, depending upon the differentiated state of the cell. Optimal quantitative control of gene expression might be better achieved in heterozygotes than in either homozygote, particularly if long stretches of DNA are involved with many differences in nucleotide composition between two homologous strands of controlling DNA. Even so it would seem that with sufficient effort, it should be possible to produce truly homozygous individuals that are economically superior to any current heterozygous individual.

When homozygosity is approached by inbreeding, several years are required to obtain reasonably homozygous strains of laboratory rodents. However, if one begins with wild animals like the field mouse, *Peromyscus,* the whole scheme of breeding usually fails after four or five generations of inbreeding. Fertility frequently declines rapidly, and the process grinds to a halt. Even when successful, the production of each homozygous strain requires thousands of animals, and it is years before one knows whether the presumptive strain will finally possess the desired characteristics. However, if homozygous diploids are made by the microsurgical procedures just described, the problem is considerably simplified. It is quite possible to enucleate 100 ova in an afternoon, and even if the success rate is only 10% in terms of young born, one could expect to produce 1000 or more homozygotes in 6 months as presumptive initiators of new strains. Perhaps 995 of them might not be worth anything, but five might be very desirable. With this procedure, it would be possible to circumvent all of the blocks of inbreeding depression and sterility and reach the end in the form of homozygosity at once. This would obviate the wasted effort of working with potential strains that were not meant to be. That is, the vast majority of unwanted genotypes will not interfere with discovering the few that are wanted.

Possibly it will be too expensive to maintain stocks by repeated enucleation. If one could microsurgically replace an X chromosome with a Y chromosome, then true strains of both males and females would be obtained. But this is beyond current technology. What can be done is to combine homozygous diploid production of females with mating mother to son and mother to grandson, etc., to produce inbred lines with both sexes more quickly than with present breeding regimens in which both males and females are heterozygous.

The low rates of development to term of homozygous diploid mice deserve additional consideration. Experience with these animals is almost exactly parallel to the general experience of investigators with parthenotes, which are also essentially homozygous when derived from inbred strains. These animals, as noted earlier, never develop to term. That result and the experiences with homozygotes (Modlinski, 1980) suggest some fundamental biological problem with homozygosity, with inbreeding. It is well-known that many biological mechanisms in many kinds of organisms operate to prevent related eggs and sperm from combining and insure that extensive outbreeding does in fact occur. Perhaps some kind of imprinting of the genome of the egg and the sperm occurs during gametogenesis such that they are different, and the difference is necessary for normal development. Recall that one of the two X chromosomes is inactivated at random in most cells of female eutherian mammals (Lyon, 1972) early in development. But in marsupials, the X chromosome that enters the ovum in the sperm is always the one inactivated (Sharman, 1971). Thus, the X chromosomes are not alike in ability to function in the fertilized egg (Cooper *et al.*, 1971). Yet in the next generation, both X chromosomes function alike in oogenesis. They are not, therefore, permanently different. The differentiation was epigenetic and reversible. In other words, the preceding cellular environment of the X chromosome makes a difference in how it performs in the newly fertilized egg and in the cells of the subsequent embryo and adult. There are some Japanese mice with genes that cause lethality on outcrossing. This varies according to whether the genes are inherited maternally or paternally (Wakasugi, 1974). The same phenomenon is also known in plants. One can extrapolate the behavior of these genes and chromosomes to the whole genome. In certain insects, for example the mealybug, the male genome is epigenetically different from the female genome, and that difference is important for normal development (Brown and Mur, 1964; Brown and Wiegmann, 1969). Similarly, complete homozygosity in mammals may create a block to normal development. However, if understood, this block might be circumvented. Such a block could explain why the parthenotes fail to develop to term and why homozygous mouse embryos generally die during development, granting that there has been one report of success.

V. CLONING BY NUCLEAR TRANSPLANT

A. Microinjection of Sperm

Microinjection of sperm, just the reverse of microsurgical removal of male pronuclei, is perhaps the simplest variant of transplanting nuclei. Potentially, it is an extremely powerful technique, even outside of the context of cloning. For example, if one were trying to make hybrids between mice and rats, injecting sperm would circumvent the main barrier to such fertilization, which is penetration of the zona pellucida by the foreign sperm. One could also use sperm of certain genotypes of animals that are incapable of normally fertilizing ova (Johnson and Hunt, 1971; Afzelius *et al.*, 1975). One could overcome the deficiencies of such sperm by separating their heads and tails by sonication (a process which also usually removes the acrosomes), picking up the heads in a pipette, and injecting them into the oocyte. There is clear evidence that rat sperm will participate in embryonic development after injection into mouse ova, since chromosomes of both species are found on the spindle that forms prior to the first mitotic division (Thadani, 1979, 1980). Such hybrid embryos probably would not go much further than the blastocyst stage *in vivo*; they have not been cultured beyond the two cell stage *in vitro*. However, it appears that the injected sperm heads function like normal sperm when injected into ova by this method. It can also be concluded that the sperm does not have to fuse with the membrane of the oocyte in order to participate in embryonic development. The fact that the karyotype of the two-cell stage has complete sets of both normal rat and mouse chromosomes suggests that perfectly normal individuals might be produced by injecting sperm into ova of the same species. Thus, one could hope to eliminate the necessity of using large numbers of sperm for fertilizing ova—one sperm per egg should be sufficient. So, if one still has a vial of frozen semen from an outstanding bull with a few million sperm in it, it might be enough to last forever. In *Xenopus*, the sperm injection technique has resulted in live normal progeny (Brun, 1974).

Injection of sperm heads may also allow testing whether all sperm are functionally equivalent or not. There is a notion that considerable selection of sperm takes place in the female reproductive tract and that many kinds of genetic or developmentally defective sperm are eliminated on the way to the site of egg fertilization (Cohen, 1973). With *in vitro* fertilization, such selection is circumvented, which leads to fears of increased abnormalities among embryos. With sperm injection, that issue could be tested directly. If the incidence of abnormalities after injection of unselected sperm is no higher than usual, then one does not need to worry about the sperm's race to the ovum; possibly every sperm is just as good as any other even though some are slower to reach the ova than others.

B. Microinjection of Somatic Cell Nuclei

Cloning a prize bull to make 50 other prize bulls in an afternoon by transplanting somatic cell nuclei would obviously be economically worthwhile. The difficulty is that it would require cloning from an adult. Cloning from embryos, in the case of fish and amphibia, is not too difficult (Chapter 9, this volume). To clone from embryonic mammals may also not be so difficult (Modlinski, 1978). Illmensee and Hoppe (1981) have recently reported success in transplanting nuclei from cells of the inner cell mass of mouse embryos into fertilized 1-cell ova. With the same pipette, they removed the two pronuclei of the fertilized egg. Following transfer to recipients, three of these clones developed into normal offspring. This is true cloning of mammals from embryonic cells, and it represents what can reasonably be expected from present technology. Unfortunately, from an economic standpoint, one does not want to clone embryos when it is unclear whether the embryo is worth cloning. Furthermore, the methods described earlier for making identical multiplets are simpler means of accomplishing the same thing. For cloning nonhomozygous adults, however, nuclear transplant remains a desirable goal.

C. Differentiation of Somatic Cell Nuclei

Unfortunately, biological evidence suggests, quite strongly, that at least some, if not most, nuclei of adult cells are irreversibly different from those in early embryos, and it is probable that the difference is genetic, i.e., the DNA has been irreversibly modified. If so, cloning from such cells would be impossible (Talmage, 1979; Sakano *et al.*, 1980). There are a number of reasons for believing that this may be the true situation. Earlier it was thought that adult cells were identical in chromosomal makeup, and in fact in a gross karyotypical sense, they clearly are. It has been too easy to extrapolate from a superficial observation to an ultimate conclusion, in this case that the chromosomal complements in the nuclei of adult cells are all identical to one another. This conclusion was bolstered by successful cloning from embryonic cells, and even from a few cells a bit farther along, as from the larvae of *Xenopus laevis* (Gurdon, 1975; Dawid and Wahli, 1979). Since cells of a developing individual could be used for cloning, a big leap occurred in accepting the idea that nuclei of an adult individual would work too, as if all nuclei in an individual were genetically identical, with exactly the same DNA unmodified from the fertilized egg. However, efforts to clone from adult amphibians have failed (Gurdon *et al.*, 1975; Briggs, 1977; McKinnell, 1978 and Chapter 9, this volume; DiBerardino, 1980). They will probably fail in mammals for exactly the same reasons—namely, that the genome differentiates irreversibly as cells mature during embryonic development. Clearly, adult cells are very stable in their phenotypic manifestations—liver cells stay as liver cells, and blood cells as blood cells, even after many cell divisions. One likely reason for this stability is that the nuclei are truly different so that differentiated cells do not change even after many types of perturbations to their cytoplasm.

Scientists have recently come to appreciate that the DNA is not quite as fixed as once thought (Abelson, 1980). For example, the formation of specific immunoglobulins was recently postulated to arise from selective loss of DNA (Talmage, 1979; Sakano *et al.*, 1980). It is also now known, for example, that DNA in different parts of a chromosome in lymphocytes contributes to transcription of a single RNA molecule. Also, once the transcripts are formed, they undergo great modification through excision of certain RNA sequences and specific resplicing. There is evidence from plants and some animals that certain genes move around from one place in the genome to another. In addition, viruses can be inserted in specific parts of the genome and can be excised again. These kinds of observations make it easy to imagine slight rearrangements in the genome as it differentiates to achieve stability for a fixed and differentiated phenotypic function. If this occurs, cloning from ordinary adult mammalian cells may be biologically impossible, although the eventual ability to cause dedifferentiation in some cases represents an intriguing research goal.

There is one cell type, though, which should be immune to these problems—spermatogonia. DiBerardino and Hoffner (1971) tested the totipotency of amphibian adult spermatogonia by transplanting the spermatogonial nuclei into enucleated amphibian ova. Normal development did not occur. Nevertheless, the total effort was not so great that we should consider the issue settled. After all, transplanting any kind of nuclei leads to a certain fraction of failures. Without any doubt, the nucleus of the spermatogonium is totipotential. The DNA cannot be irreversibly differentiated because it will give rise to a sperm, which in turn can give rise to a complete individual. If one accepts the report of Hoppe and Illmensee (1977), a single sperm nucleus actually did give rise to a complete animal in two instances. Spermatogonia have several other advantages; they are dividing diploid cells and only a small biopsy from the testes is required to obtain them. There is no particular mechanical problem in removing the nucleus and injecting it into an ovum, so there is a reasonable hope that spermatogonia might be used successfully to clone males. Clearly, spermatogonia would be the logical choice for cloning with adult cells. With the kind of technology described previously for enucleation of mouse ova, it is obviously possible to work with mammalian ova in the same way that it has been possible to work with amphibian ova.

VI. BEYOND CLONING

Although making identical genetic copies of outstanding animals has numerous uses and advantages, clearly there are also problems (Table I). Even if it were possible to carry out these procedures easily and inexpensively, they by no means represent the ultimate in genetic engineering. Eventually, one should be able to produce genotypes to specifications that optimize the end products desired, taking into account the environment in which the animal will be placed. Biologists can obviously obtain the ova, manipulate them in various ways, even take chromosomes out and put other chromosomes in, culture the embryos, transfer them back into recipient females, and in some cases the embryos will develop to term. Thus, the basic technology for such manipulation of the ova and sperm of mammals is at hand (Table II), though the biological problems just described clearly represent constraints. Full fruition of the technology, e.g., sperm injection, requires a substantial investment of effort, resources, and time,

TABLE I

Cloning for Animal Industry

Uses and advantages
  1. Removal of genetic variation in experiments
  2. Rapid production of inbred lines
  3. Copying small numbers of animals with outstanding genotypes for breeding
  4. Making many copies of animals with outstanding phenotypes for production
  5. Rapid production of lines with special traits (e.g., twinning in cattle)
  6. Automatic sex selection

Disadvantages
  1. Probable low rates of success
  2. Probably cumbersome and expensive
  3. Problems of embryo culture and transfer
  4. May be detrimental to donors of nuclei
  5. Some methods result in females only
  6. Difficulties in working with differentiated nuclei
  7. Potential for decreased genetic variation in populations
  8. May upset economics of purebred industry

TABLE II

Manipulations of Genetic Instructions in Oocytes

---

1. Normal fertilization
2. Dispermy, one male (selfing); removal of female pronucleus
3. Dispermy, two males (crossing); removal of female pronucleus
4. Fusion of two oocytes from one female (selfing)
5. Fusion of two oocytes from different females (crossing)
6. Diploid parthenogenesis (consequences same as 4 above)
7. Homozygous diploid, female pronucleus
8. Homozygous diploid, male pronucleus
9. Haploid parthenogenesis, cytochalasin B (consequences same as 7 above)
10. Injection of somatic cell nucleus, removal of pronuclei
11. Production of identical twins

---

but the investment would probably pay off handsomely, perhaps even in ways not even dreamed of today.

REFERENCES

Abelson, J. (1980). *Science (Washington, D.C.)* **209**, 1319-1321.

Afzelius, B. A., Eliasson, R., Johnsen, Ø., and Lindholmer, C. (1975). *J. Cell Biol.* **66**, 225-232.

Beatty, R. A. (1967). "Fertilization" (C. Metz and A. Monroy, eds.), Vol. 1, pp. 413-440. Academic Press, New York.

Bedford, J. M., and Overstreet, J. W. (1972). *J. Reprod. Fertil.* **31**, 407-414.

Briggs, R. (1977). *In* "Cell Interactions in Differentiation" (M. Karkinen-Jääskeläiner, L. Saxen, and L. Weiss, eds.), pp. 23-43. Academic Press, New York.

Brown, S. S., and Spedich, J. A. (1979). *J. Cell Biol.* **83**, 657-662.

Brown, S. W., and Mur, U. (1964). *Science (Washington, D.C.)* **145**, 130-136.

Brown, S. W., and Wiegmann, L. (1969). *Chromosoma* **28**, 255-279.

Brun, R. B. (1974). *Biol. Reprod.* **11**, 513-518.

Cohen, J. (1973). *Heredity* **31**, 408-413.

Cooper, D. W., VandeBerg, J. L., Sharman, G. B., and Poole, W. E. (1971). *Nature (London) New Biol.* **230**, 155-157.

Cuellar, O. (1971). *J. Morphol.* **133**, 139-165.

Cuellar, O. (1977). *Science (Washington, D.C.)* **197**, 837-843.

Dawid, I. B., and Wahli, W. (1979). *Dev. Biol.* **69**, 305-328.

DeFord, L. S., Buss, E. G., Todd, P., and Wood, J. C. S. (1979). *J. Exp. Zool.* **210**, 301-306.

DiBerardino, M. A. (1980). *Differentiation (Berlin)* **17**, 17-30.

DiBerardino, M. A., and Hoffner, N. (1971). *J. Exp. Zool.* **176**, 61-72.

Edwards, R. G. (1957). *Proc. R. Soc. London Ser. B* **146**, 488-504.

Eppig, J. J., Kozak, L. P., Eicher, E. M., and Stevens, L. C. (1977). *Nature (London)* **269**, 517-518.

Fernandez, M. S., and Izquierdo, L. (1980). *Anat. Embryol.* **160**, 77-81.

Gerritsen, J. (1980). *Am. Nat.* **115**, 718-742.

Gillespie, L. L., and Armstrong, J. B. (1979). *J. Exp. Zool.* **210**, 117-121.

Ginzburg, L. R. (1979). *Theor. Popn. Biol.* **15**, 264-267.

Graham, C. F. (1974). *Biol. Rev.* **49**, 399-422.

Gurdon, J. B. (1975). *In* "Cell Cycle and Differentiation" (J. Reinert and H. Holtzer, eds.), pp. 123-131. Springer-Verlag, Berlin and New York.

Gurdon, J. B., Laskey, R. A., and Reeves, O. R. (1975). *J. Embryol. Exp. Morphol.* **34**, 93-112.

Hertwig, O. (1911). *Arch. Mikrosk. Anat. Abt. II* **77**, 1-164.

Hoppe, P. C., and Illmensee, K. (1977). *Proc. Natl. Acad. Sci. U.S.A.* **74**, 5657-5661.

Hsu, Y. C., and Gonda, M. A. (1980). *Science (Washington, D.C.)* **209**, 605-606.

Illmensee, K., and Hoppe, P. C. (1981). *Cell* **23**, 9-18.

Johansson, I., Lindhé, B., and Pirchner, F. (1974). *Hereditas* **78**, 201-234.

Johnson, D. R., and Hunt, D. M. (1971). *J. Embryol. Exp. Morphol.* **25**, 223-226.

Kaplan, R., Dekel, N., and Kraicer, P. F. (1978). *Gamete Res.* **1**, 59-63.

Kaufman, M. H , and Gardner, R. L. (1974). *J. Embryol. Exp. Morphol.* **31**, 635-642.

Kaufman, M. H., and O'Shea, K. S. (1978). *Nature (London)* **276**, 707-708.

Kaufman, M. H., Barton, S. C., Azim, M., and Surani, H. (1977). *Nature (London)* **265**, 53-55.

Lyon, M. F. (1972). *Biol. Rev.* **47**, 1-35.

McKinnell, R. G. (1978). "Cloning, Nuclear Transplantation in Amphibia." Univ. of Minnesota Press, Minneapolis.

Markert, C. L., and Petters, R. M. (1977) *J. Exp. Zool.* **201**, 295-302.

Maynard Smith, J. (1978). "The Evolution of Sex." Cambridge Univ. Press, London and New York.

Modlinski, J. A. (1978). *Nature (London)* **273**, 466-467.

Modlinski, J. (1980). *J. Embryol. Exp. Morphol.* **60**, 153-161.

Moustafa, L. A., and Hahn, J. (1978). *Dtsch. Tieraerztl. Wochenschr.* **85**, 242-244.

Mukai, T., Chigura, S., and Yoshikawa, I. (1964). *Genetics* **50**, 711-715.

Mullen, R. J., Whitten, W. K., and Carter, S. C. (1970). *In* "Annual Report of the Jackson Laboratory," pp. 67-68. Bar Harbor, Maine.

Patterson, J. T. (1913). *J. Morphol.* **24**, 559-684.

Reinschmidt, D. C., Simon, S. J., Volpe, E. P., and Tompkins, R. (1979). *J. Exp. Zool.* **210**, 137-143.

Rugh, R. (1939). *Proc. Am. Philos. Soc.* **81**, 447-471.

Sakano, H., Maki, R., Kurosawa, Y., Roeder, W., and Tonegawa, S. (1980). *Nature (London)* **286**, 676-683.

Seidel, F. (1952). *Naturwissenschaften* **39**, 355-356.

Seidel, G. E., Jr. (1980). *Proc. 9th Int. Congr. Anim. Reprod. Artif. Insem., Madrid,* **2**, 363-370.

Sharman, G. B. (1971). *Nature (London)* **230**, 231-232.

Snow, M. H. L. (1973). *Nature (London)* **244**, 513-515.

Sorensen, R. A. (1973). *Am. J. Anat.* **136**, 265-276.

Soupart, P., Anderson, M. L., and Repp, J. E. (1978). *Theriogenology* **9**, 102 (Abstr.).

Stevens, L. C. (1978). *Nature (London)* **276**, 266-267.

Stevens, L. C., Varnum, D. S., and Eicher, E. M. (1977). *Nature (London)* **269**, 515-517.

Steinhardt, R. A., Epel, D., Carroll, E. S., and Yanagimachi, R. (1974). *Nature (London)* **252**, 41-43.

Suomalainen, E. (1950). *Adv. Genet.* **3**, 193-253.

Surani, M. A. H., Barton, S. C., and Kaufman, M. H. (1977). *Nature (London)* **270**, 601-603.

Talmage, D. W. (1979). *Am. Sci.* **67**, 173-177.

Tarkowski, A. K., and Wroblewska, J. (1967). *J. Embryol. Exp. Morphol.* **18**, 155-180.

Thadani, V. T. (1979). *J. Exp. Zool.* **210**, 161-168.

Thadani, V. T. (1980). *J. Exp. Zool.* **212**, 435-453.

Thadani, V. T. (1981). Ph.D. Thesis, 362 pp. Yale Univ., New Haven, Conn.

Tompkins, R. (1978). *J. Exp. Zool.* **203**, 251-255.

Tsukahara, J., and Ishikawa, M. (1980). *Eur. J. Cell Biol.* **21**, 288–295.

Uzzell, T. (1970). *Am. Nat.* **104**, 433–445.

Wakasugi, N. (1974). *J. Reprod. Fertil.* **41**, 85–96.

Willadsen, S. M. (1979). *Nature (London)* **277**, 298–300.

Willadsen, S. M. (1980). *J. Reprod. Fertil.* **59**, 357–362.

Willadsen, S. M., Lehn-Jensen, H., Fehilly, C. B., and Newcomb, R. (1981). *Theriogenology* **15**, 23–29.

11

Gene Transfer in Mammalian Cells

DAVOR SOLTER

I. INTRODUCTION

Gene transfer from one organism to another resulting in new and unique combinations of genetic material occurs with every fertilization. Randomness inherited in this process, which can be only minimally controlled by breeding programs, precludes the use of this system of gene transfer in studying gene control and function. It is therefore not surprising that scientists attempted to devise methods by which gene transfer could be better controlled.

201

NEW TECHNOLOGIES IN ANIMAL BREEDING

New techniques for this control, developed within the last 5 years, enable us for the first time to envision genetic engineering as a practical proposition. The possible practical, ethical, and philosophical impact of genetic engineering on society forces us to examine as objectively as possible the present and potential applications of this new technology.

Experimental approaches for the gene transfer between mammalian cells started 20 years ago with introduction of somatic cell hybridization. However, this technique and its modifications accomplished only random gene transfer. Even so, somatic cell hybridization provided a great deal of information about gene mapping, gene function, and gene control. Techniques developed within the last 5 years encompassed under the term *recombinant DNA technology* now permit the introduction of a selected gene into a given cell and the study of how this gene is controlled. Recent results show that controlled introduction of foreign genes into whole organisms will be feasible, raising both the level of application of the technique and the level of possible studies of gene action.

The purpose of this article is to briefly describe current techniques used in gene transfer and the types of scientific questions posed and answered as a result. Only the barest outline and a guide for further reading is offered, since the subject of gene transfer is enormous.

II. RANDOM AND PARTIALLY DIRECTED GENE TRANSFER

A. Somatic Cell Hybridization

Somatic cell hybridization assumes the fusion (spontaneous or experimentally induced) between, usually, two different cells. The immediate result of such a fusion is a *heterokaryon*, a single cell containing both parental nuclei, but a random mixture of other cellular components from parental cells. A heterokaryon can progress through mitosis during which the parental nuclear material mixes together. The resulting cells are *somatic cell hybrids,* which have a single nucleus composed of randomly selected chromosomes derived from both parental cells. A somatic cell hybrid may and usually does have an unstable genotype and, during subsequent divisions, may lose some chromosomal material. With time, chromosome loss becomes negligible, and the hybrids develop into a stable cell line in which all cells are identical and possess a unique chromosomal set derived from chromosomes of both parental cells. Thus, a single fusion event can result in numerous different (chromosomally) somatic cell hybrid lines.

1. Fusion of Whole Cells

Several discoveries made somatic cell hybridization one of the most powerful techniques in genetics today; detailed accounts of these results have been published (for further readings, see Ephrussi, 1972; Davidson and de la Cruz, 1974; Harris, 1974; Ringertz and Savage, 1976). The basic technique used today (shown schematically in Fig. 1) is the result of several important advances that made somatic cell hybridization easy and reliable. The first somatic cell hybrids (Barski *et al.*, 1960) were the result of spontaneous fusion and were observed because they grew faster than either of the parental cells. Spontaneous fusion is a very rare event, and the resulting hybrids usually grow much slower than parental cells. It was therefore necessary to improve the method of hybrid selection and increase the efficiency and incidence of fusion. The first (and most widely used) method of selection was introduced by Littlefield (1964). It is based on the selection of mutant cells defective in specific enzymes by growing cells in the

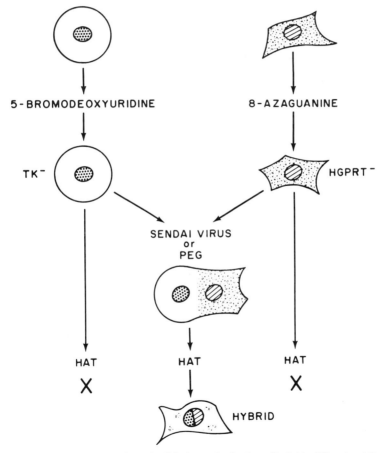

Fig. 1. Schematic presentation of cell fusion and selection of hybrids. TK⁻, thymidine kinase deficient cell; HGPRT⁻, hypoxanthine guanine phosphoribosyltransferase-deficient cells; PEG, polyethylene glycol; HAT, selective medium containing hypoxanthine, aminopterin, and thymidine.

presence of drugs (Szybalski *et al.*, 1962). Cells grown in the presence of bromodeoxyuridine (BRdU) die because the enzyme thymidine kinase (TK) incorporates the drug into DNA; cells lacking this enzyme (TK⁻) survive. Similarly, cells grown in the presence of azaguanine die because the enzyme hypoxanthine guanine phosphoribosyltransferase (HGPRT) converts the drug into metabolites that interfere with nucleic acid synthesis. Again, HGPRT⁻ cells survive. Absence of these enyzmes does not interfere with cell growth in normal medium, since they are involved only in the salvage pathways of nucleotide synthesis. However, if TK⁻ and HGPRT⁻ cells are grown in medium containing hypoxanthine, aminopterin, and thymidine (HAT medium developed by Szybalski *et al.*, 1962), they die. The major synthetic pathway is blocked by aminopterin, and the cells cannot use hypoxanthine and thymidine as precursors for nucleotide synthesis. Thus, if TK⁻ cells are fused with HGPRT⁻ cells and placed in HAT medium, all cells that did not fuse and cells that fused with the same type of cells die. Only the hybrids between TK⁻ and the HGPRT⁻ cells will survive as each parent contributes one of the enzymes necessary for survival. By introducing HAT selective medium, Littlefield (1964) made possible the reliable and easy selection of even very few hybrid cells amidst the large number of unfused parental cells.

Spontaneous fusion as mentioned earlier is extremely rare. Observation that inactivated Sendai virus induces cell fusion *in vitro* (Okada, 1958) led to the use of Sendai virus in production of interspecific heterokaryons (Okada and Murayama, 1965; Harris and Watkins, 1965). Inactivated Sendai virus is now the most commonly used agent, although others have been described, most notably polyethylene glycol (PEG) (Pontecorvo, 1976).

Somatic cell hybridization was used in the ensuing years to study several very important biological questions: expression of differential gene function in hybrids of different phenotype, control of malignancy in hybrids between malignant and normal cells, control of virus infectivity and expression in hybrids between permissive and nonpermissive cells, activation of dormant nuclei, and so on. The finding that in interspecific hybrids (between cells from two different species) chromosomes of one parent are preferentially lost (Weiss and Green, 1967) enabled construction of hybrids containing only one or few chromosomes from one parental cell. Combined with precise chromosome identification, the use of such hybrids made possible mapping of numerous human genes and advanced tremendously formal human genetics.

Somatic cell hybridization has recently led to the development of another field of enormous practical and scientific importance, namely hybridoma technology. Köhler and Milstein (1975) first demonstrated that hybrids between mouse myeloma cells and spleen cells from a mouse specifically immunized will continuously produce antibodies specific for the antigen used. These hybrids producing monoclonal antibodies offer an unlimited supply of the same well-defined antibody and make possible the precise characterization of antigenic molecules. The availability of monoclonal antibodies will certainly facilitate existing immunodiagnostic methods and introduce new ones. Monoclonal antibodies have been successfully used in immunodetection (Ballou *et al.*, 1979) and immunotherapy (Bernstein *et al.*, 1980) of mouse tumors. Recently developed human monoclonal antibodies produced either by mouse–human (Schlom *et al.*, 1980) or by human–human hybrids (Olsson and Kaplan, 1980; Croce *et al.*, 1980) herald the introduction of monoclonal antibodies that can be invaluable tools in immunodetection and immunotherapy of human tumors.

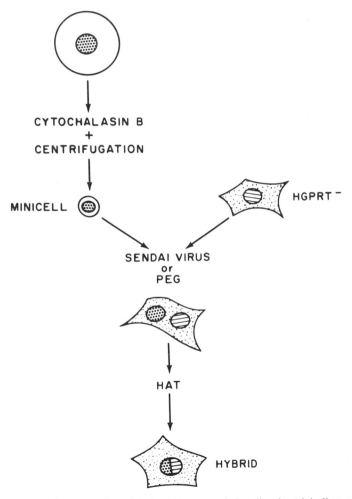

Fig. 2. Schematic presentation of fusion between a whole cell and a minicell or nucleoplast. Abbreviations as in Fig. 1.

2. Nucleoplasts, Cytoplasts, and Cell Reconstruction

An alternative to fusion of whole cells was developed a few years ago. This made possible the fusion of cellular components. The method was based on the original observation by Carter (1967) that cells treated with cytochalasin B extrude nuclei; this method is depicted in Fig. 2. Cytochalasin B causes outpocketing of nuclei within a small cytoplasmic pouch attached to the main body of the cell by a cytoplasmic stalk. Such stalks can be broken by centrifugation resulting in the formation of minicells or nucleoplasts, which are composed of a nucleus surrounded by a small rim of cytoplasm and a plasma membrane. The remaining body of the cell without the nucleus is called a *cytoplast*. Both nucleoplasts and cytoplasts are viable for some time, and since they are surrounded by plasma membrane, they can be fused with other cells (for details, see Ege *et al.*, 1974; Veomett *et al.*, 1974; Bossart *et al.*, 1975). The use of enucleation procedures offered some new possibilities in somatic cell hybridization. From a technical standpoint, the limited life-span of nucleoplasts eliminated the time-consuming procedure of selecting for drug-resistant mutants, and, thus, any cell could then be used for fusion. Nucleoplasts from one cell line can be fused with cytoplasts from another. We can then start asking questions about the effect of the cytoplasm on the phenotypic character of the cell, the role of cytoplasmic components in control of gene expression, and so on.

Somatic cell hybridization using whole cells or nucleoplasts still resulted in mixing of complete genetic material from two different cell lines. Subsequent sorting out and the resulting hybrids were essentially random products. Attention will now be turned to attempts to limit the genetic contribution of the parent.

B. Transfer of Chromosomes

1. Microcells and Liposome Carriers

The first method designed to introduce one or a few chromosomes into other cells is an extension of the method used for fusion of nucleoplasts (Fig. 3). By exposing the cells to colcemid, the cells are arrested in mitosis and chromosomes are condensed. Subsequent treatment with cytochalasin and centrifugation produces microcells composed of only a few chromosomes, a small amount of cytoplasm, and a plasma membrane. Microcells can be fused with other cells, and the selection is the same as described for minicells (for details, see Ege and Ringertz, 1974; Fournier and Ruddle, 1977).

Chromosomes can also be transferred using liposome carriers instead of minicells (Fig. 4). Metaphase chromosomes are isolated from cells treated with colcemid and subsequently disrupted. The chromosomes are then trapped in lipid vesicles, and such lipochromosomes are fused with normal cells using PEG (Mukherjee *et al.*, 1978). Hybrids are then selected as described previously.

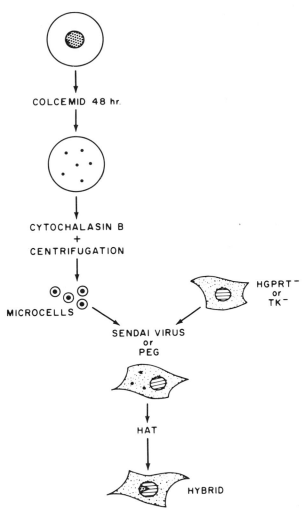

Fig. 3. Schematic presentation of fusion between whole cell and microcell. Microcells contain only one or few chromosomes. Abbreviations as in Fig. 1.

Incidentally, liposomes and similar types of vesicles are being used extensively for introducing foreign molecules into cells. Possibilities of using such carriers to deliver specific molecules (drugs) to a particular cell type in the body open up fascinating approaches to specific drug targeting (for details, see Celis *et al.*, 1979; Baserga *et al.* 1980; Poste, 1980). The use of monoclonal antibodies in combination with liposomes makes precise targeting even more feasible (Huang *et al.*, 1980; Leserman *et al.*, 1980).

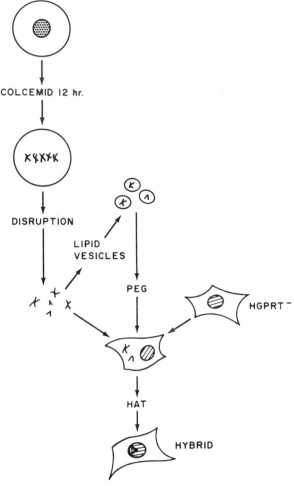

Fig. 4. Schematic presentation of introduction of foreign genetic material in cells either by chromosomes entrapped in lipid vesicles (lipochromosomes) or by naked metaphase chromosomes. Abbreviations as in Fig. 1.

2. Chromosome and DNA-Mediated Gene Transfer

This chapter has to this point dealt with the transfer of genetic material using as a vehicle the whole cell, part of the cell, or an artifically constructed vesicle. Possibilities of transferring genes by exposing cells directly either to isolated chromosomes or purified DNA have also been explored for some time. As early as 1971, Merril *et al.* reported successful transfer of lambda phage DNA into mammalian cells. Subsequently McBride and Ozer (1973) reported the transfer of genetic information by isolated mammalian metaphase chromosomes into other mammalian cells. The method involves exposing cells to colcemid and disrupting these mitotically arrested cells and collecting the chromosomes. Recipient cells are exposed to isolated chromosomes and grown in selective medium (Fig. 4). Subsequently numerous gene transfers between several species were reported (Burch and McBride, 1975; Spandios and Siminovitch, 1978; Klobutcher and Ruddle, 1979; Deisseroth and Hendrick, 1979). Using isolated metaphase chromosomes for gene transfer, Wigler *et al.* (1980) reported the transformation of mammalian cells using purified and fragmented genomic DNA.

It should be emphasized that all the methods described so far involved some kind of selective pressure to isolate the cells that received the genetic material. Consequently, the only genes that were definitely transferred were those that insured the survival of the hybrid. With somatic cell hybrids, a large number of other genes were transferred and retained by the hybrid, whereas transformation with metaphase chromosomes or DNA resulted in hybrids that had no identifiable foreign chromosomes and probably only a small amount of transferred genetic material (McBride *et al.*, 1978). The selective marker was obviously always present and in addition other genes close to the selective markers were detected. These results indicate that metaphase chromosomes upon entering the cells underwent fragmentation and that the small fragments integrated into the chromosomes of the host. Thus, it became possible to introduce a small number of genes into cells. The particular genes transferred depended on the selective system used. Transfer of nonselectable genes was entirely due to their closeness to the selectable markers and was not controlled by the experimentation. Techniques developed for the introduction of either selectable or nonselectable genes into cells will now be discussed.

III. DIRECTED GENE TRANSFER

A. Isolation and Amplification of Genes—Recombinant DNA Technology

In order to transfer a particular gene from one organism to another, the gene must first be available in purified form and in sufficient quantity. Until recently this seemed next to impossible, but the discovery of several important enzymes and the introduction of new techniques greatly simplified these tasks. Only a few essential steps in recombinant DNA technology will be discussed here. Details can be found elsewhere (Scott and Werner, 1977; Wu, 1979; Grossman and Moldave, 1980; Abelson and Butz, 1980).

In order to clone a gene, the first step is to devise a probe that can detect the

210   **Davor Solter**

gene in its genomic form. The most common and simplest forms of cloning involve the following steps (Fig. 5).

1. Isolation of messenger RNA (mRNA) on oligo-dT columns. mRNA is polyadenlated at the 3′ end (AAA) and remains in the column.

2. Isolated mRNA is then transcribed using the enzyme reverse transcriptase (isolated from RNA tumor viruses) and oligo-dT primer into an RNA-DNA hybrid.

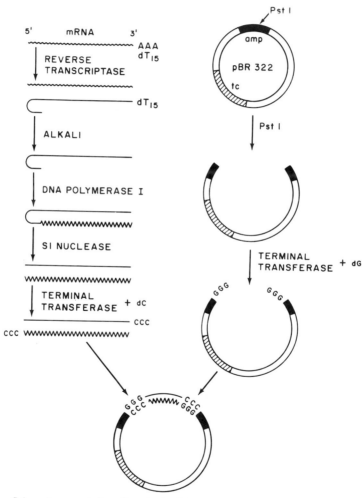

Fig. 5. Schematic presentation of synthesis of double-stranded DNA (ds-cDNA) on messenger RNA (mRNA) template and incorporation of ds-cDNA into plasmid pBR322. A, adenine; dT, deoxythymidine; dC, deoxycytidine; dG, deoxyguanine; *amp*, ampicillin resistance site in the plasmid; *tc*, tetracycline resistance site; Pst 1, site where endonuclease Pst 1 cuts plasmid DNA.

3. After alkali treatment to separate the strands of RNA and DNA, the second strand of DNA is synthesized by DNA polymerase I.

4. Treatment with S1 nuclease removes hairpin loop and single strands at the end of DNA molecules, and a terminal transferase adds cytosine (CCC) residues.

5. A plasmid (pBR322) is cut with endonuclease Pst 1 in the ampicillin (*amp*) resistance site, and a terminal transferase adds guanosine (GGG) residues.

6. In the mixture of double-stranded DNA (ds-cDNA) and linearized plasmid, the guanosine and cytosine residues pair, and the recombinant plasmid containing the cloned piece of DNA is formed.

7. Insertion of DNA into the *amp* site destroys ampicillin resistance, whereas tetracycline (*tc*) resistance is maintained. Bacteria infected with such plasmids grow in the presence of tetracycline but not in the presence of ampicillin, and this selective system can be used to isolate the desired plasmid clones. The cloned piece of DNA can always be isolated from the plasmid by using endonuclease Pst 1 again, since the recombinant plasmid was constructed in such a way as to form Pst 1 sites at both ends of the inserted fragment.

8. Using ds-cDNA cloned from mRNA, the isolation and cloning of the corresponding genomic DNA can be achieved. Although cDNA probes are often used to isolate corresponding genomic sequences, there are several other methods that have been used to isolate genomic sequences of interest (Scott and Werner, 1977; Setlow and Hollaender, 1979).

Assuming successful isolation and cloning of the genes, one must now consider the methods of introducing them into cells. First attempts to transfer a particular cloned gene into cells used the same selective systems described before, the model system being the introduction of the purified viral thymidine kinase gene into TK⁻ cells (Wigler *et al.*, 1977). Transformants (cells with integrated viral gene) were then isolated in HAT selective medium. Only a small proportion of the cells become transformed, but (analogous to the work done in bacteria) cells that are transformed with one gene are likely to be transformed with others as well. Thus, simultaneous exposure of these cells to the thymidine kinase gene and other purified but not selectable genes allowed the introduction of the desired nonselectable genes into mammalian cells (for example, see Wigler *et al.*, 1979; Pellicer *et al.*, 1980). Direct microinjection of purified genes might increase the efficiency of transformation (Anderson *et al.*, 1980).

B. Viruses and Genes as Vectors for Other Genes

In order to improve the efficiency of introducing nonselectable genes into cells, these genes can be linked in plasmids to either selectable genes or to viruses. Using the viral thymidine kinase gene as a vector and HAT selection medium (Fig. 6), several genes such as the rabbit β-globin gene (Mantei *et al.*, 1979) and the chicken ovalbumin gene (Lai *et al.*, 1980) have been introduced into mouse cells. Viruses or viral genes, which can either permanently replicate

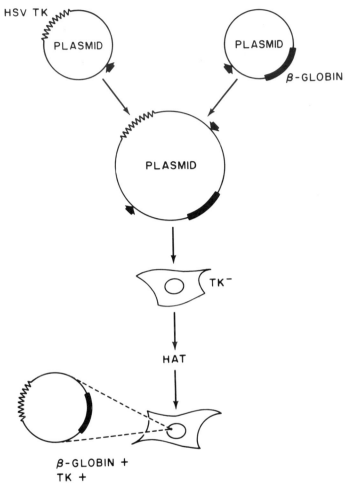

Fig. 6. Schematic presentation of introduction of β-globin gene into cell using enzyme thymidine kinase as vector. *HSV TK*, herpes simplex virus thymidine kinase gene (other abbreviations as in Fig. 1).

in infected cells or transform normal cells, have also been used (Fig. 7) to introduce the mouse β-globin into monkey cells with the resulting production of the mouse protein (Hamer *et al.*, 1979; Hamer and Leder, 1979). Mouse teratocarcinoma cells are resistant to transformation with SV40 virus, but the viral genome was successfully integrated into TK⁻ teratocarcinoma cells using a plasmid containing the SV40 genome and the thymidine kinase gene (Linnenbach *et al.*, 1980). This sytem was then used in analyzing the control of viral expression in differentiating cells (Knowles *et al.*, 1980).

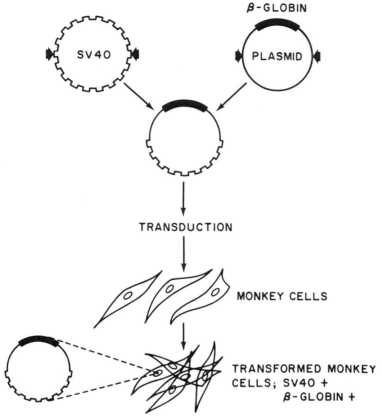

Fig. 7. Schematic presentation of introduction of β-globin gene into cell using Simian virus 40 (SV40) as vector.

The primary purpose of gene transfer is to study the control of gene expression. Many complex phenomena may be observed after successful transfer. Questions surrounding chromosomal localization of the transferred gene, transfer of controlling elements along with the coding sequences, and the possibility of *in vitro* altering of the gene sequence before transfer (Scherer and Davis, 1979) are beyond the scope of this article. Some interesting accounts of problems involving gene transfer have recently been published (Marx, 1980).

IV. ORGANISMIC VERSUS CELLULAR GENE TRANSFER

To completely understand the function of a particular gene and the mechanisms involved in controlling its expression, it is essential to observe the consequences of its introduction not only in cells *in vitro* but also in normal tissues and whole organisms. If genetic therapy is ever to become a reality, one must be able to introduce genes into organisms in such a way that they do not disturb the function of other genes and that they ensure the proper expression and control of these introduced genes. Some possible approaches to this problem are projected in the following sections.

A. Introduction of Foreign Genes into Tissue

The experimental strategy of introducing foreign genes into tissue is theoretically simple. Normal donor cells are treated *in vitro* so that desired gene(s) are introduced. These cells are then placed into a recipient where they continue to grow and proliferate expressing the introduced genes. These cells must possess some selective advantage over corresponding cells of the recipient in order to assure their survival and eventual overgrowth. For example, bone marrow cells of mice were made methotrexate-resistant by introducing the increased number of copies of genes coding for dihydrofolate reductase. These cells were then injected into mice that were irradiated to eliminate endogenous bone marrow cells and were treated with methotrexate to provide a selective advantage to transferred cells. Subsequently, cells carrying the introduced genes colonized the recipients' bone marrow (Cline· *et al.*, 1980; Mercola *et al.*, 1980). The same group of scientists have recently attempted similar experiments in human patients, but the results are as yet unknown.

B. Introduction of Foreign Genes into Whole Organisms

The only way that genes can be introduced into all cells of the organism is to introduce them either in fertilized eggs or in very young embryos. So far, the only successful experiment of this kind was reported by Jaenisch (1977), who injected a mouse RNA tumor virus into a mouse blastocyst and found that the germ cells in some adults derived from injected blastocyst had integrated viral genome. Upon breeding those adults, the viral genome was transmitted to all the cells of their progeny.

Several laboratories have attempted to introduce foreign genes by directly injecting purified genes into the nucleus of the mouse zygote. There are as yet only unpublished reports that viral DNA was detectable at birth in a small number of mice derived from zygotes injected with SV40 genes. These types of experiments will certainly continue with increased vigor.

One problem with injecting genes into zygotes is the total absence of a selective system. Introduced genes do not confer any advantage to the developing embryo, and if transformation efficiency is low, large numbers of embryos would have to be manipulated in order to produce some with integrated genes.

One possible strategy to avoid this problem is depicted in Fig. 8. This experimental protocol is largely hypothetical, but there are indications that it might work. Embryonal carcinoma cells (ECC) are the stem cells of a tumor called *teratocarcinoma* (Sherman and Solter, 1975). ECC resemble embryonic cells in most of their characteristics (Solter and Damjanov, 1979; Martin, 1980). Most importantly, ECC can participate in the formation of chimeric animals when injected into normal mouse blastocysts (Brinster, 1974; Illmensee and Mintz, 1976), which suggests that ECC are actually normal embryonic cells growing as tumors

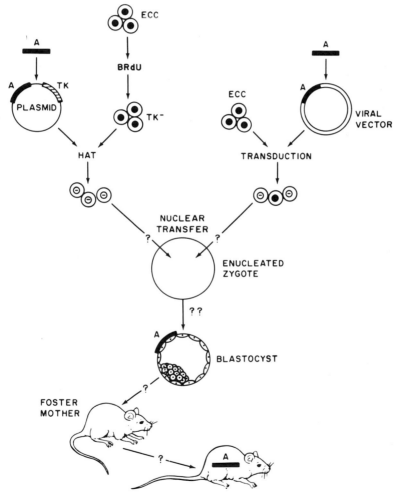

Fig. 8. Hypothetical protocol for transfer of gene A into whole animal. ECC, embryonal carcinoma cell; BRdU, bromodeoxyuridine (other abbreviations as in Fig. 1).

only in an extrauterine environment. When such cells are injected into a blastocyst, they express a normal developmental potential. ECC can be manipulated *in vitro* so that foreign genes are stably introduced (Linnenbach *et al.*, 1980; Pellicer *et al.*, 1980), and the resulting cells are not obviously different from the original populations. There is no reason to think that the methods for introducing genes into other cell types would not work with ECC as well.

Another set of experiments relevant to this protocol has recently been performed by Illmensee and Hoppe (1981). They introduced the nuclei from embryonic cells into enucleated zygotes and obtained live mice. This first report of successful nuclear transfer and cloning in mammals has broad scientific and practical application, but its importance for the present discussion lies in the possibility that nuclei from ECC might eventually also be used for cloning. Several tremendous obstacles still exist, the most important being the requirement for complete euploidy of ECC since chromosomal aberrations are usually incompatible with normal development. However, if ECC nuclei are able to support the development of enucleated zygotes into adult animals, our ability to introduce isolated and purified genes into organisms will be virtually limitless.

V. CONCLUDING REMARKS

Several methods have been described for accomplishing genetic transfer from one cell to another and from one individual to another. Each of these methods allows a new insight into complex problems of the expression and regulation of mammalian genes, and it can be predicted with confidence that the possibilities are by no means exhausted. Introduction of these methods also drastically altered the practical applicability of basic biology and genetics and, also, changed society's perception of these sciences. Production of monoclonal antibodies, hormones, and other biologically active molecules is transforming genetic research from an academic pursuit into an important national resource. Scientists can only hope that the upheavals that follow any announcement of a new technology will subside. Rational assessments of the benefits of new genetic technologies should prevail over the superstitious panic that has become almost a reflex reaction of one very vocal part of our society. Any scientific discovery has a potential for misuse, but only level-headed approaches and complete understanding of discoveries will insure the proper role of science in our future.

ACKNOWLEDGMENTS

The author apologizes to numerous scientific colleagues whose work was not properly acknowledged due to the lack of space. Also, I am aware that oversimplifications and misinterpretations in the text might have occurred for the same reason. Special thanks are given to my colleagues, Barbara B.

Knowles and Chris Overton, for critically reading the manuscript. During the writing of this manuscript, support was received from U.S. Public Health Service Research Grants CA-10815 and CA-27932 from the National Cancer Institute, by HD-12487 from the NICHHD, and by Grant PCM 78-16177 from the National Science Foundation.

REFERENCES

Abelson, J., and Butz, E., eds. (1980). *Science (Washington, D.C.)* **209**, 1319-1438.

Anderson, W. F., Killos, L., Sanders-Haigh, L., Kretschmer, P. J., and Diacumakos, E. G. (1980). *Proc. Natl. Acad. Sci. U.S.A.* **77**, 5399-5403.

Ballou, B., Levine, G., Hakala, T. R., and Solter, D. (1979). *Science (Washington, D.C.)* **206**, 844-847.

Barski, G., Sorieul, S., and Cornefert, F. (1960). *C.R. Acad. Sci.* **251**, 1825-1827.

Baserga, R., Croce, C., and Rovera, G., eds. (1980). "Introduction of Macromolecules into Viable Mammalian Cells." Alan R. Liss, New York.

Bernstein, I. D., Tam, M. R., and Nowinski, R. C. (1980). *Science (Washington, D.C.)* **207**, 68-71.

Bossart, W., Loeffler, H., and Bienz, K. (1975). *Exp. Cell Res.* **96**, 360-366.

Brinster, R. L. (1974). *J. Exp. Med.* **140**, 1049-1056.

Burch, J. W., and McBride, O. W. (1975). *Proc. Natl. Acad. Sci. U.S.A.* **72**, 1797-1801.

Carter, S. B. (1967). *Nature (London)* **213**, 261-266.

Celis, J. E., Graessmann, A., and Loyter, A., eds. (1979). "Transfer of Cell Constituents into Eukaryotic Cells." Plenum, New York.

Cline, M. J., Stang, H., Mercola, K., Morse, L., Ruprecht, R., Browne, J., and Salser, W. (1980). *Nature (London)* **284**, 422-425.

Croce, C. M., Linnenbach, A., Hall, W., Steplewski, Z., and Koprowski, H. (1980). *Nature (London)* **288**, 488-489.

Davidson, R. L., and de la Cruz, F., eds. (1974). "Somatic Cell Hybridization." Raven, New York.

Deisseroth, A., and Hendrick, D. (1979). *Proc. Natl. Acad. Sci. U.S.A.* **76**, 2185-2189.

Ege, T., and Ringertz, N. R. (1974). *Exp. Cell Res.* **87**, 378-382.

Ege, T., Krondahl, U., and Ringertz, N. R. (1974). *Exp. Cell Res.* **88**, 428-432.

Ephrussi, B. (1972). "Hybridization of Somatic Cells." Princeton Univ. Press, Princeton, New Jersey.

Fournier, R. E. K., and Ruddle, F. H. (1977). *Proc. Natl. Acad. Sci. U.S.A.* **74**, 319-323.

Grossman, L., and Moldave, K., eds. (1980). "Nucleic Acids," Part I, Academic Press, New York.

Hamer, D. H., and Leder, P. (1979). *Nature (London)* **281**, 35-40.

Hamer, D. H., Smith, K. D., Boyer, S. H., and Leder, P. (1979). *Cell* **17**, 725-735.

Harris, H. (1974). "Nucleus and Cytoplasm." Oxford Univ. Press, London and New York.

Harris, H., and Watkins, J. F. (1965). *Nature (London)* **205**, 640-646.

Huang, A., Huang, L., and Kennel, S. J. (1980). *J. Biol. Chem.* **255**, 8015-8018.

Illmensee, K., and Hoppe, P. C. (1981). *Cell* **23**, 9-18.

Illmensee, K., and Mintz, B. (1976). *Proc. Natl. Acad. Sci. U.S.A.* **73**, 549-553.

Jaenisch, R. (1977). *Cell* **12**, 691-696.

Klobutcher, L. A., and Ruddle, F. H. (1979). *Nature (London)* **280**, 657-660.

Knowles, B. B., Pan, S., Solter, D., Linnenbach, A., Croce, C., and Huebner, K. (1980). *Nature (London)* **288**, 615-618.

Köhler, G., and Milstein, C. (1975). *Nature (London)* **256**, 495-497.

Lai, E. C., Woo, S. L. C., Bordelon-Riser, M. E., Fraser, T. H., and O'Malley, B. W. (1980). *Proc. Natl. Acad. Sci. U.S.A.* **77**, 244-248.

Leserman, L. D., Barbet, J., Kourilsky, F., and Weinstein, J. N. (1980). *Nature (London)* **288,** 602–604.

Linnenbach, A., Huebner, K., and Croce, C. M. (1980). *Proc. Natl. Acad. Sci. U.S.A.* **77,** 4875–4879.

Littlefield, J. W. (1964). *Science (Washington, D.C.)* **145,** 709–710.

McBride, O. W., and Ozer, H. L. (1973). *Proc. Natl. Acad. Sci. U.S.A.* **70,** 1258–1262.

McBride, O. W., Burch, J. W., and Ruddle, F. H. (1978). *Proc. Natl. Acad. Sci. U.S.A.* **75,** 914–918.

Mantei, N., Boll, W., and Weissmann, C. (1979). *Nature (London)* **281,** 40–46.

Martin, G. R. (1980). *Science (Washington, D.C.)* **209,** 768–776.

Marx, J. L. (1980). *Science (Washington, D.C.)* **210,** 1334–1336.

Mercola, K. E., Stang, H. D., Browne, J., Salser, W., and Cline, M. J. (1980). *Science (Washington, D.C.)* **208,** 1033–1035.

Merril, C. R., Geier, M. R., and Petricciani, J. C. (1971). *Nature (London)* **233,** 398–400.

Mukherjee, A. B., Orloff, S., Butler, J. B., Triche, T., Lalley, P., and Schulman, J. D. (1978). *Proc. Natl. Acad. Sci. U.S.A.* **75,** 1361–1365.

Okada, Y. (1958). *Biken J.* **1,** 103–110.

Okada, Y., and Murayama, F. (1965). *Exp. Cell Res.* **40,** 154–158.

Olsson, L., and Kaplan, H. S. (1980). *Proc. Natl. Acad. Sci. U.S.A.* **77,** 5429–5431.

Pellicer, A., Wagner, E. F., El Kareh, A., Dewey, M. J., Reuser, A. J., Silverstein, S., Axel, R., and Mintz, B. (1980). *Proc. Natl. Acad. Sci. U.S.A.* **77,** 2098–2102.

Pontecorvo, G. (1976). *Somatic Cell Genet.* **1,** 397–400.

Poste, G. (1980). *In* "Liposomes in Biological Systems" (G. Gregoriadis and A. C. Allison, eds.), pp. 101–151. Wiley, New York.

Ringertz, N. R., and Savage, R. E. (1976). "Cell Hybrids." Academic Press, New York.

Scherer, S., and Davis, R. W. (1979). *Proc. Natl. Acad. Sci. U.S.A.* **76,** 4951–4955.

Schlom, J., Wunderlich, D., and Teramoto, Y. A. (1980). *Proc. Natl. Acad. Sci. U.S.A.* **77,** 6841–6845.

Scott, W. A., and Werner, R., eds. (1977). "Molecular Cloning of Recombinant DNA." Academic Press, New York.

Setlow, J. K., and Hollaender, A., eds. (1979). "Genetic Engineering, Principles and Methods." Plenum, New York.

Sherman, M. I., and Solter, D., eds. (1975). "Teratoma and Differentiation." Academic Press, New York.

Solter, D., and Damjanov, I. (1979). *Methods Cancer Res.* **18,** 277–332.

Spandios, D. A., and Siminovitch, L. (1978). *Nature (London)* **271,** 259–261.

Szybalski, W., Szybalska, F. H., and Ragni, G. (1962). *Natl. Cancer Inst. Monogr.* **7,** 75–89.

Veomett, G., Prescott, D. M., Shay, J., and Porter, K. R. (1974). *Proc. Natl. Acad. Sci. U.S.A.* **71,** 1999–2002.

Weiss, M. C., and Green, H. (1967). *Proc. Natl. Acad. Sci. U.S.A.* **58,** 1104–1111.

Wigler, M., Silverstein, S., Lee, L. S., Pellicer, A., Cheng, Y. C., and Axel, R. (1977). *Cell* **11,** 223–232.

Wigler, M., Sweet, R., Sim, G. K., Wold, B., Pellicer, A., Lacy, E., Maniatis, T., Silverstein, S., and Axel, R. (1979). *Cell* **16,** 777–785.

Wigler, M., Perucho, M., Kurtz, D., Dana, S., Pellicer, A., Axel, R., and Silverstein, S. (1980). *Proc. Natl. Acad. Sci. U.S.A.* **77,** 3567–3570.

Wu, R., ed. (1979). "Recombinant DNA." Academic Press, New York.

IV

Analysis of Impact of Technology on Animal Breeding

12

Potential Genetic Impact of Artificial Insemination, Sex Selection, Embryo Transfer, Cloning, and Selfing in Dairy Cattle

L. DALE VAN VLECK

221

NEW TECHNOLOGIES IN ANIMAL BREEDING
Copyright © 1981 by Academic Press, Inc.
All rights of reproduction in any form reserved.
ISBN 0-12-123450-9

I. INTRODUCTION

Recent advances in reproductive biology appear to promise rapid genetic improvement in farm livestock. Improved genetic potential generally means decreased relative cost to the consumer, because as each livestock unit produces more, the maintenance cost of the unit is a smaller portion of the product. This discussion will be limited to one class of livestock—dairy cattle, to one trait—milk production, and to an outline of the principles to be considered in economic and genetic analyses of the potential of these man-made alterations in the mode of distributing genetic material of animals designated as superior. Other reports, which have considered portions of this topic and other classes of livestock, are listed in the References (Foote and Miller, 1971; Meadows, 1974; Skjervold, 1974; Cunningham, 1975; Land and Hill, 1975; Van Vleck, 1975, 1976; Hill and Land, 1976; Kräusslich, 1976; Hansen, 1976; Petersen and Hansen, 1977; Wilmut and Hume, 1978).

A difficult decision to make in any such analysis is who should be the beneficiary of the improvement. In a controlled society, the general population is the obvious and logical choice, but individuals and organizations are the basic economic units, so perhaps the beneficiary of both the gain and the risk should be the economic units. Decisions also must be made on the basis of long-term or short-term gains. In general, society would be interested in long-term gains, whereas individuals may be interested in either or both. In this presentation, the long-term gain for population and the short-term gain for individual dairymen will be considered. *Long-term gain* will therefore be defined as the gain in the population that accumulates over a relatively long period. *Short-term gain* will be defined as the gain for an individual dairyman over a definite period with the period including value of production gains and possibly the sale of animals for a specific number of years. Any economic genetic analysis must define the beneficiary and the time period to be considered.

II. BASIC GENETIC AND ECONOMIC PRINCIPLES

There are three genetic equations that can be used as a basis for predicting either short- or long-term genetic gain. Portions of these equations can be altered by the mode of reproduction, and, thus, the equations allow comparisons of various breeding systems. An assumption that is commonly made is that the records and genetic values follow a multivariate normal distribution. This assumption allows use of normal theory in predicting gain from selection. A second, although less tenable assumption, is that genetic variation and normality are maintained through several generations of selection.

A. Superiority of Selected Group

The expected superiority in additive genetic value of the selected group above that of those available for selection is the well-known

$$\Delta G = r_{G\hat{G}} I \sigma_G \tag{1}$$

where $r_{G\hat{G}}$ is the correlation between the actual and predicted additive genetic value, also called accuracy of evaluation, I is the standard normal selection differential (Table I), and σ_G is the standard deviation of additive genetic value. The standard deviation is fixed, but the other two terms can be altered by the breeding systems.

B. Genetic Gain per Year

Since male and female selection may be different (different $r_{G\hat{G}}$ and I), long-term genetic gain per year can be approximated by the formula of Rendel and

TABLE I

Standardized Normal Selection Factors

Fraction selected	Factor	Fraction selected	Factor	Fraction selected	Factor	Fraction selected	Factor
0.001	3.400	0.01	2.660	0.10	1.755	0.55	0.720
0.002	3.200	0.02	2.420	0.15	1.554	0.60	0.644
0.003	3.033	0.03	2.270	0.20	1.400	0.65	0.570
0.004	2.975	0.04	2.153	0.25	1.271	0.70	0.497
0.005	2.900	0.05	2.064	0.30	1.159	0.75	0.424
0.006	2.850	0.06	1.985	0.35	1.058	0.80	0.350
0.007	2.800	0.07	1.919	0.40	.966	0.85	0.274
0.008	2.738	0.08	1.858	0.45	.880	0.90	0.195
0.009	2.706	0.09	1.806	0.50	.798	0.95	0.109

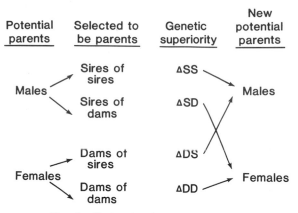

Fig. 1. Four paths of genetic improvement.

Robertson (1950) extended from that of Dickerson and Hazel (1944), which shows the contributions of four paths (Fig. 1) to genetic gain:

$$\Delta g = \frac{\Delta SS + \Delta SD + \Delta DS + \Delta DD}{L_{SS} + L_{SD} + L_{DS} + L_{DD}}$$

where ΔSS is the superiority of males selected to be sires of sires; ΔSD is the superiority of males selected to be sires of dams; ΔDS is the superiority of females selected to be dams of sires; ΔDD is the superiority of females selected to be dams of dams; and L is the corresponding generation interval. Thus, the generation interval is a factor in addition to $r_{G\hat{G}}$ and I, which will be affected by the breeding system. Various modifications of this formula have been developed (Specht and McGilliard, 1960; Searle, 1961; Skjervold, 1963, 1966; Skjervold and Langhols, 1964; Van Vleck, 1964; Bichard, 1971; Bichard et al., 1973; Brascamp, 1973; Hunt et al., 1974; Hill, 1974), but for most purposes this formulation is probably sufficient. In the application of this equation, the assumption will be made that the population size is large enough so that choices can be made about such factors as number of bulls to test, number of daughters per bull, number of bulls used in proved bull studs, and number of dams of sons, among other factors. For example, a breed with only 25,000 required first services would have difficulty in testing 50 bulls per year with 500 first services each and still be able to select the best for heavy use.

C. Progeny Superiority

The third equation can be used to predict short-term gain and results from the fact that progeny will receive a sample one-half of genes from both parents.

Thus, the expected superiority of progeny from selected parents above that of randomly mating at a particular time is

$$\Delta P = \frac{\Delta S + \Delta D}{2}$$

where ΔS is $r_{G_S \hat{G}_S} I_S \sigma_G$, the expected superiority of the selected males and ΔD is $r_{G_D \hat{G}_D} I_D \sigma_G$, the expected superiority of the selected females.

If actual predicted values of the sires and dams are available, then those averages can be substituted into the equation to predict progeny superiority.

D. Inbreeding

Another genetic factor, which may need to be considered, is inbreeding. Some of these genetic delivery systems can markedly reduce the number of males and females needed to produce each new generation. The increase in inbreeding is dependent on the population size—number of males and females used for breeding. The usual formula for increase in inbreeding per generation is:

$$\Delta F = (N_S + N_D)/(8 \times N_S N_D)$$

where N_S and N_D are the number of sires and dams needed per generation. The general effects of inbreeding in reduction of viability and production are known, although not very precisely.

Most inbreeding experiments with animals have resulted in eventual failure of the lines to reproduce. Whether this difficulty can be overcome with some of the new reproductive procedures, which can increase inbreeding very rapidly, is unknown. In any case, inbreeding may result in a cost in terms of lowered viability and production. There are also two other related implications of increased homozygosity. Quantitative genetic principles can in theory handle the changed relationships resulting from inbreeding, but in practice there are no adequate corrections for inbreeding depression in doing evaluations for additive genetic effects. In addition, lifetime production is a function also of viability, which may be substantially affected by inbreeding, and of capacity to produce if the animal lives, i.e., productivity = prob (survival) × expected (production/survival). Thus, two traits—survival and production—are involved in a nonlinear way, which also introduces some problems of theory, although the quadratic index of Wilton et al. (1968) would seem to apply. One final difficulty in projecting gain over several generations is that inbreeding should diminish, as should selection, genetic differences and thus the genetic standard deviation.

E. Discounted Genetic Gains

Costs other than from inbreeding must also be considered. Each breeding system will have different costs. These costs as well as income from genetic gains should be discounted (interest and inflation rate considered) for the time period under evaluation. Gene flow to future generations, survival probabilities, and actual rather than age-adjusted production should also be considered. The procedure of Everett (1975) seems a reasonable method for dairy production. Hill (1974), Bichard (1971), and Bichard et al. (1973) have considered gene flow for multiple birth species. The procedure of Everett was used by Van Vleck and Everett (1976) to establish the value of sexed semen. In fact, the procedure predicts the net present value of any mating after considering discount rate, investment period, survival probabilities to various lactations, predicted records adjusted to age at lactation, and descendents' production, given that the mating results in a freshening heifer. Reproductive costs to obtain the heifer must be deducted to establish the predicted net present return over investment. This would be the amount of money that, if invested at the time of mating at the corresponding interest rate and for the corresponding investment period, would be expected to yield the same return as the mating. The mathematics involves survival matrices but results in table values (Table II) which can be easily used to predict returns from matings. The table value times predicted genetic value of the heifer times the net price per pound of milk (milk price minus feed cost) gives the net present value relative to an animal with zero genetic value. In many cases, it can be used to predict the differences in returns between two kinds of matings or within herd selection procedures. Examples will be given in discussion of sexing

TABLE II

Fraction of a Heifer's Genetic Value Expected in Production of the Heifer and Her Descendents Adjusted for Survival Probabilities and Age of Production and Discounted Back to Time of Her Conception for Four Investment Periods and Eight Discount Rates

Years in invest-ment period	True discount rate							
	0.00	0.02	0.04	0.06	0.08	0.10	0.12	0.14
5	2.20	2.04	1.88	1.74	1.62	1.50	1.40	1.30
10	5.32	4.70	4.18	3.72	3.33	2.99	2.69	2.44
15	7.16	6.13	5.29	4.60	4.02	3.53	3.13	2.79
20	8.31	6.94	5.87	5.01	4.31	3.74	3.28	2.90

Value of mating which produces heifer[a,b] = (table value) $\left(\dfrac{\hat{G}_{sire}}{2} + \dfrac{\hat{G}_{dam}}{2} \right)$ (net value/lb milk)

[a] Assumes heifer calf born and freshens.

[b] For comparing selection of dams, the sire part may be assumed to be the same or zero.

semen and embryo transfer, which may allow more intense selection (and higher genetic values) of the dams of the heifers.

Many of the genetic delivery systems can work together and most can interact with artificial insemination, which will be discussed briefly.

III. ARTIFICIAL INSEMINATION

The two-stage selection used by most artificial insemination (AI) studs for dairy cattle incorporates (a) mating of the most superior dairy bulls to the most superior cows based on predicted genetic values, followed by (b) random mating of their sons through AI to obtain a progeny evaluation and subsequent culling of the poorest sons. This has been very effective. The development of the theory of optimum AI breeding systems includes work of Robertson and Rendel (1950), Henderson and Dunbar (1952), Henderson (1954, 1964), Searle (1961), Van Vleck (1964), Skjervold and Langholz (1964), and Skjervold (1963, 1966), among others. Optimum plans under natural within herd breeding systems could at best provide a gain of 0.5–0.6%/year. Optimum use of AI in large populations could increase this to 2.0–2.5% to, perhaps, 3% per year. The reasons are the increased accuracy of evaluation, together with more intense selection of sires of dams, sires of sires, and dams of sires. Female selection does not become much more accurate with AI and selection of dams of dams also cannot be much more intense. Generation intervals could be reduced somewhat through AI since progeny of sires can be obtained over a shorter period of time than with natural service. Inbreeding has not become much of a problem with most breeds because of the large population of cows and the many bulls used, although not necessarily recommended. An example of the calculations for predicting progress from AI is shown in Table III. Average accuracy values with AI are shown in the top row of the table. Easily obtainable intensities of selection are in the second row of the table for an AI situation. The expected yearly gain would be 220 lb of milk per cow. The maximum expected annual gain from natural use of bulls (NS) would be about 50 lb per cow. The entry on the far right in the table gives the expected net present value for cows selected to leave heifer replacements in the herd. The value under an AI program will serve as a basis for comparison for sexed semen and embryo transfer plans. Gain from use of AI bulls over the past 20 years has been substantial (Van Vleck, 1977) and has been much more rapid than gain from NS bulls, but has fallen far short of 220 lb of milk.

AI organizations have not rigorously applied the basic principles of selection. They have not practiced strong enough selection for sires of bulls and cows and for dams of bulls because of reluctance to break with traditional selection practices, because of financial constraints for proper testing of young bulls to produce sires of cows, and because of too much emphasis put on nonproductive traits of

TABLE III

Example of Influence of Sexing Semen and Embryo Transfer on the Four Paths for Genetic Gain in Milk Yield per Year

	Path					Net present value
Assumed accuracy	SS 0.79	SD 0.79	DS 0.65	DD 0.65	Gain/year[a]	of mating to selected DD[b]
Regular AI						
Selection factor	2.153	1.400	1.985	.195	220 lb	$14
Percent selected	4	20	6	90		(79 lb)[c]
Sexed semen						
Selection factor	2.153	1.400	2.270	.880	253 lb	$63
Percent selected	4	20	3	45		(358 lb)
AI with embryo transfer						
Selection factor	2.153	1.400	2.660	1.755	296 lb	$126
Percent selected	4	20	1	10		(713 lb)
Embryo transfer—few bulls						
Selection factor	2.420	2.420	2.660	1.755	349 lb	$126
Percent selected	2	2	1	10		(713 lb)
Embryo transfer and sexed semen—few bulls						
Selection factor	2.420	2.420	2.900	2.064	367 lb	$148
Percent selected	2	2	0.5	5		(838 lb)

[a] Genetic standard deviation/sum of generation intervals = 1250/24 = 52.08.
[b] For genetic superiority in brackets and discount rate of 0.10, 15-year investment period and $0.05 net return per pound of milk (Table II value = 3.53).
[c] Genetic superiority of heifers due to selection of dams.

questionable economic value (Vinson, 1975; Van Vleck, 1977). Accuracy of evaluation of dams of bulls has also been less than theoretically predicted (Van Vleck, 1969; Van Vleck and Carter, 1972; Butcher and Legates, 1976). Only about 33% of that predicted superiority has been achieved, probably because dairymen can by conscious effort make cows with good early production continue to produce better than they would with random treatment. This factor alone could account for 40–45 lb of the decrease from expected gain. Methods of evaluating the genetic value of cows that circumvent the natural desire of the dairymen to make a good cow seem better are needed. The gain to the industry can be substantial.

Methods to optimize economic gain from AI mating plans have led to more and more sophisticated discounting procedures (Searle, 1961; Van Vleck, 1964; Hill, 1971; Hinks, 1971; Brascamp, 1973; Vaccaro, 1974). Enormous strides also have been made in obtaining accurate genetic evaluations of bulls from records of their female relatives by the use of procedures developed by Henderson (1973, 1975). The problems and potential of the application of AI in dairy cattle breeding were recently reviewed by Van Vleck (1977).

IV. SEXED SEMEN

Sexing of semen to produce heifer calves for dairymen has been attempted without significant success for many years. If successful, sex control would likely impact mostly on intensity of selection. Foote and Miller (1971) discussed the impact on the population with varying degrees of successful sex control and for all classes of farm livestock. Van Vleck and Everett (1976) used the net present value procedure described earlier to calculate the economic value of sex control for the buyer of sexed semen. Their conclusion based on application of Eq. (3) and Table II to predict discounted production was that, for a 5-year investment period, the value of sexed semen over regular semen would be about $11 per ampule and, for a 20-year period, the value of sexed semen would be about $22 more than regular semen. Sale of cull calves was assumed to be the same for sexed and regular semen systems. Semen used on cows not selected to produce replacements was assumed to have a minimum price. The expected small increase in value of sexed semen for the dairyman is in spite of a greatly increased selection differential on dams of dams—from the top 90% with $I_{0.90} = 0.195$ as compared to the top 45% with $I_{0.45} = 0.880$. The genetic value of semen to produce replacement heifers was assumed to be equal for both systems. An example using more current milk prices (net of 0.05/lb), a 0.10 discount rate, and a 15-year period is shown on the right side of Table III.

The marginal cost of sex-controlled semen can be calculated approximately for this example. The usual AI program requires about six services to produce a heifer that freshens in the herd. At $10 per service, the cost would be $60. With sexed semen, only the one-half of the cows selected to produce replacement heifers would be bred with sexed semen. The other one-half would be bred to less expensive semen. Thus, on the average, three services with sexed semen and three services with other semen would be required to yield a fresh heifer. Assume the inexpensive service is $7 and the sexed semen service is $10 + $SS. The cost would be $51 + 3$SS. The return over cost with regular AI (Table III) is $14 − $60 = −$46. The return over cost with sexed semen (Table III) is $63 − $51 − 3$SS. To have equal return, $63 − $51 − 3$SS = −$46. The cost of sexing the semen would need to be less than $58/3 = $19 to be profitable. The cost of sexing the semen must consider any losses of semen in the sexing process.

The increase in genetic gain due to selection of dams of dams would be nearly fivefold, 29.8 lb/year versus 6.6 lb/year, with a total sum of generation intervals of 24 years with sexed semen as compared to AI alone. Similar results were predicted by Foote and Miller (1971).

The gain for the population considering all four paths could be somewhat more. Again, perfect sexing will be assumed in the rough example (see Table III). The other three paths are SS, SD, and DS. Only one-half as many dams of sons would be needed. The increase in I depends on the fraction currently needed. If the top 6% are now used, $I_{0.06}$ is 1.985, and if only 3% are needed, then $I_{0.03}$ is 2.27 for an increase in that path of about 14%, with the assumption of equal accuracy of evaluating dams of sons. This increase in intensity of selection for dams of sons could be quite easily met under current conditions. In that case, sexed semen could still halve the number of dams of sons needed, but the increase in the intensity factor would be relative less—2.27 to 2.53, or 11%.

Without detailed examination, the net effect of sex control on the sire paths appears to be little. If only heifers are desired, then only one-half as much semen would be available, but only one-half as many cows are needed to produce replacement heifers. The other one-half of the cows could be bred with relatively inexpensive semen or with the other one-half of the sperm, which produces bull calves. Loss of semen in the sexing process might require additional sires and force a corresponding decrease in the selection factor. Enough semen is readily available from superior sires to produce sons so that no increase in selection intensity would seem likely. Accuracy of evaluation would also seem to be little changed since about the same fraction of the cow population would be used for evaluating bulls. The example shows the theoretical genetic gain to be 33 lb more per year for sexed semen as compared to AI alone of which 23 lb is due to selection of dams of replacement heifers.

Academic animal breeders would argue that 1% of cows rather than 6% is sufficient for selecting dams of bulls. In that case, sexed semen would provide opportunity to select the top 0.5%. Gain from the dam of bull path with sexed semen would be 31 lb, making the total expected gain 274 lb/year versus 220 lb/year for AI alone.

The value of added gain of 30 lb of milk per year is not great for an individual cow, but when multiplied by a national herd of 7,000,000 cows and accumulated for 10 years, the economic value would be large at 10¢/lb—about 1 billion dollars, an average of 100 million per year and 210 million in year 10 alone.

The cost of sexing semen is not known since no one has been successful. Probably if a way is found, the cost will not be great but will need to be under $10 to $20 per breeding unit to be economical.

Sex control can work together with embryo transfer to increase further the selection differentials. The main potential of embryo transfer is to increase selection intensity and probably not to increase accuracy of evaluation.

V. EMBRYO TRANSFER

Embryo transfer has had its major use in obtaining more than the normal number of offspring from highly prized cows or cows of highly priced breeds. The goal was a high sale value (Cunningham, 1976). The long-term goal, however, should be improved genetic value. Individual dairymen then would be interested in whether gain from being able to select replacement heifers from, e.g., the top 10% of the herd, by use of embryo transfer, rather than the top 90% would pay for the added costs. Some crude calculations by Van Vleck (1975) indicated that for a 10-year investment period, combined yearly embryo transfer and semen costs could not exceed $1070 per 100 cow herd more than usual semen costs in order for embryo transfer to be profitable, i.e., about $10.70 per cow in the herd.

Another approach (Van Vleck, 1976) is to compare the expected net present value from herd use of embryo transfer from top cows to produce replacement heifers with regular selection (intensity of 1 of 10 versus 9 of 10). Table II gives the discounted value of 1 lb of superiority in genetic ability per milking heifer for various time periods and discount rates. The table value is multiplied by expected genetic superiority of the heifers and net return per pound of milk. To predict net present value, reproductive costs of producing a fresh heifer should then be deducted when comparing breeding systems.

Assume an accuracy of 0.65 for cows and a genetic standard deviation of 1250 lb. Selecting from the top 90 of 100 gives $I_{0.90} = 0.195$ and $\Delta G = (0.195) (0.65) (1250) = 158$ of which one-half goes to the heifer (AI row of Table III).

The discounted net present value at $0.05/lb would be at 0.10 discount rate and for 15 years: $(3.53) (0.05) (79) = \$14$. If replacements were selected from the top 1 of 10 with embryo transfer, $I_{0.10} = 1.755$ and the discounted net present value would be $(3.53) (0.05) (713) = \$126$. These values are given in the right most column of Table III.

The reproductive cost of obtaining a fresh heifer would be about six inseminations for regular breeding and about 2.4 live calves for embryo transfer (conception rate of 50%, chance of live birth 80%, sex ratio of 50%, and survival rate after birth of 83%). This crude analysis would indicate that each successful live birth by embryo transfer would be worth about ($126 − $14)/2.4 = $47, with some adjustment still required for amount of semen used. The true discount rate can have a substantial influence on predicted return on investment. At 0.00% rate, the net present values would be about twice those for 0.10%, and the added value of a live birth by embryo transfer would increase to about $95.

The calculations in the previous paragraph were done assuming the same accuracy of evaluating cows, since by the time a cow has several daughters with records by embryo transfer she will have had so many records herself that accuracy would increase only slightly. The intensity of selection, however, for the sire of bull and sire of cow paths may or may not be increased. A common practice is to superovulate the donor cow and use several times the normal quantity of semen to insure fertilization. With that practice, just as many bulls would be needed as without embryo transfer. If only a single dose is used per service, then much less semen would be needed. Fewer bulls would be needed, and if the same number as before were tested, selection could be more intense. Therefore, calculations were also made for that possibility with the prospect of 10 calves per year per selected donor cow. The calculations for the example are summarized in Table III as well as for superimposing sexing of semen on embryo transfer.

If gain from embryo transfer comes only from the dam paths, the expected gain over AI alone is 76 lb per year. Extra gain at $0.05/lb above feed cost would need to accumulate for 79 years before added gain in that year would equal a $300 embryo transfer cost per cow and for 316 years for a cost of $1200. Current popular articles indicate a potential minimum cost of $300 with nonsurgical methods, although $1200 is a more commonly quoted price. Embryo transfer in all years before balance of cost and return would yield a loss.

If less semen is needed, which would allow more intense bull selection, the expected added gain of 129 lb per year would need to accumulate for 46 years to balance embryo transfer costs of $300 per cow in the same year. Embryo transfer and perfect sexing of semen would combine to improve gain slightly.

The general conclusion is that costs of embryo transfer would need to be reduced by a large multiple in order to be economically feasible if only genetic gain is considered.

Another factor, which should be considered in a cost analysis, is that embryo transfer on a population basis might drastically reduce semen sales depending on optimum number of sperm for insemination of superovulated cows. Per ampule semen costs might rise as a result. Even more importantly, less incentive might exist for proving bulls so that the intensity factors listed in Table III would be difficult to achieve or would go in the wrong direction.

The individual dairyman, however, would be interested in the last column of

Table III since the discounted net present values given there would correspond to the genetic gains that could be made within a herd by jointly applying embryo transfer and sexing of semen to allow much more intense selection of cows to produce replacement heifers. The expected gain from joint use of sexed semen and embryo transfer as compared to embryo transfer alone is not large but should not be very much more costly than embryo transfer alone.

Embryo transfer seems unlikely to lead to increased accuracy of predicting genetic value (Meadows, 1974; Cunningham, 1976; Hill and Land, 1976). With less semen needed, fewer daughters per tested bull might result and actually decrease accuracy of evaluation. Obtaining records from progeny of a cow resulting from embryo transfer would increase the accuracy of evaluation somewhat (Table IV), but that small increase would be more than offset by an increase of 3–4 years in generation interval if selection were to be dependent on records of the progeny.

TABLE IV

Accuracy of Evaluating a Cow Using Records on Her Clones, on Her Paternal Sibs, and on Her Embryo Transferred Full-Sib Progeny[a]

	Number of clones					
Records on cow	0	1	2	3	4	5
0	0.00	0.50	0.63	0.71	0.75	0.79
1	0.50	0.63	0.71	0.75	0.79	0.82
2	0.58	0.67	0.73	0.77	0.80	0.83
3	0.61	0.69	0.75	0.78	0.81	0.83

	Number of paternal sibs					
Records on cow	0	5	10	20	50	100
0	0.00	0.25	0.32	0.38	0.44	0.47
1	0.50	0.53	0.55	0.58	0.60	0.62
2	0.58	0.60	0.62	0.63	0.65	0.66
3	0.61	0.63	0.64	0.66	0.68	0.68

	Number of full-sib progeny					
Records on cow	0	1	2	3	4	5
0	0.00	0.25	0.33	0.39	0.43	0.46
1	0.50	0.53	0.57	0.58	0.60	0.61
2	0.58	0.60	0.62	0.64	0.65	0.66
3	0.61	0.63	0.65	0.66	0.67	0.68

[a] Heritability, 0.25; repeatability, 0.50.

VI. CLONING

Each clone is genetically an exact duplicate and naturally occurs only in pairs of identical twins. Records of clones of cows could be used to evaluate accurately a cow (and her clones). Accuracy of evaluation is high, however, even if only the cow's records are used. Thus, cloning cannot be justified because of cost as a way to improve accuracy of evaluating cows.

A. Expected Gain

Potential usefulness of cloning appears to rely on finding outstanding cows and, then, mass producing them. Before mass production, records on a number of clones for testing purposes should be obtained, which may take several years. (Freezing and tissue culture techniques appear to be needed to maintain the genetic material while testing proceeds.) The reason for testing is that apparently outstanding cows may be environmentally induced freaks, which are really not greatly superior, if at all. If normal theory does hold, the expected genetic superiority would be only about 25% of the superiority of the records. This superiority can be substantial for a cow that produces 50,000 lb of milk while her herdmates produce 20,000 lb. One-fourth, or 7500 lb, of the superiority would be expected to be genetic. Optimum AI programs would require 20-30 years to achieve that much progress. Nevertheless, AI could be expected to have an advantage after that time. What is really unknown is whether normal theory holds for such large superiorities (some 12 phenotypic standard deviations). One additional danger is that genetic diversity could be substantially reduced if only a very few cows provided all clones. There would be a danger of producing cows that could not adapt to a changing world, e.g., to a complete pasture and hay diet. The history of hybrid seed corn, which gives much the same end result as cloning, provides examples of the potential difficulty of having only limited genetic variability. Still another difficulty is that after the original big boost by cloning, further selection would be impossible or, at best, difficult since all animals would be eventually the same.

The costs of cloning would be great since embryo transfer techniques would be required. In fact, 6000 lb of milk superiority at $0.05 net/lb would be needed to balance a $300 embryo transfer cost. If cows were found that met that minimum requirement, ways must be found for further improvement. A final drawback is that cloning may be impossible from mature tissues. There does not appear to be any economic justification for freezing juvenile tissue of thousands of cows in the hope that one may prove to be a genetically superior cow.

There are some implications for improving accuracy of genetic evaluation through use of clones. The cost of the increased accuracy would need to be balanced against increased genetic gain by selection due to the increased accuracy (the appendix gives the derivation of the accuracy values).

B. Accuracy of Evaluating Cows

Clones, which are, in theory, exact genetic duplicates, can provide added information in genetic evaluation, and if enough clones have records, accuracy of evaluation would approach 100%. The accuracy would apply to all the clones as well as the original clone. When heritability is 0.25 and repeatability is 0.50 (milk production) with n records on one of the animals (probably the original) and single records on m clones, the accuracy of evaluation becomes

$$[(mn + 2m + 3n)/(mn + 2m + 2n + 5)]^{1/2}$$

With only single records on m clones, this reduces to $[m/(m + 3)]^{1/2}$ as compared to $0.71 [n/(n + 1)]^{1/2}$ with n records only on a cow. Numerical values are given in Table IV and shown the typical pattern of increased accuracy—the increase is less for each added clone. The middle of Table IV gives corresponding accuracy values when the proof of the sire of the cow is used in her evaluation. Her paternal half-sibs or her progeny do not give nearly as much information about the cow as do her duplicates. Although the accuracy is increased somewhat with clones, the selection intensity factor would be decreased since fewer genetically distinct groups would be available for selection. An optimum balance would have to be found, which may or may not be more effective than current selection procedures.

Assignment of clones as mates in evaluating bulls will be discussed later when selfing of bulls is considered.

C. Inbred Clones

Another possibility with cloning techniques would be to create completely inbred animals that could then be cloned. At this point, some unanswered questions arise. Would any completely inbred animals survive? Any lethal recessive would be homozygous and cause death. If, as has been predicted with humans, each animal is carrying an average of five such bits of genetic junk, then the probability of an animal not carrying any is quite small. The chance that a particular inbred clone would not have any recessive lethals, given that the original had five, would depend on the method of arriving at the exact duplication of chromosomal material. If random gametes are duplicated, then the probability that a clone carries no recessive lethals is $(1/2)^n$, where n is the number present in the original in the heterozygous state. Thus, if $n = 2$, on the average 1 in 4 inbred clones would not be homozygous for a lethal, but if $n = 5$, the chance would be only 1 in 32. The process of screening out the lethals may not be impossible, but it certainly would be initially costly.

The other difficulty with inbreeding as mentioned earlier is in genetic evaluation if nonlethal depression occurs. Quantitative genetic theory adequately accounts for inbreeding only in terms of genes in common. Hybrid corn provides an example of the difficulty. The inbred lines used to produce either the single-cross or double-cross hybrids are not themselves productive and essentially can be evaluated only from their crosses. This problem would be apparent in the use of completely inbred female clones to produce and test mate to evaluate bulls.

VII. SELFING

An intriguing idea is to use the next best thing to cloning to evaluate bulls. Whether this is possible or not is for the biologists to answer. The assumption will be made that females can be produced that combine two random X-bearing gametes of a male. Such a procedure would be equivalent to selfing in plants. The apparent advantage would be that the genetic material of the bull could be tested for production without the usual 50% dilution from his mates. The relationships involved are somewhat novel for an animal breeder to consider. The bull is both mother and father of the progeny in a genetic sense, and the additive relationship to his progeny is 100%. The progeny are related also to each other by 100% rather than the usual 25% as paternal half-sibs. Thus far everything looks fine for evaluating a bull from his female "selfs." The difficulty is that the progeny are 50% inbred. Recessive lethals and inbreeding depression are likely to create problems. The consequence is that any simply inherited deleterious characters would show up, and bulls carrying such genes could be culled. The chance that any bulls exist that do not carry deleterious recessive genes is probably quite low. The other problem is that nonlethal effects of inbreeding complicate genetic evaluation since there are no correction factors available for effects of inbreeding. The theoretical increase in accuracy, if the inbreeding problems could be solved, must be balanced against the cost of "selfing." The goal of most testing programs is about 50 half-sib daughters for an accuracy of 0.88. Records of only 12 "selfs" would achieve the same accuracy. The "self," however, would require the costly process of embryo transfer, whereas AI organizations have little difficulty in obtaining 50 half-sib daughters.

If no problems due to inbreeding are created, the accuracy of evaluating a bull from m of his "selfs" is

$$\{m/[m+(2-h^2)/2h^2]\}^{1/2} = [m/(m+3.5)]^{1/2}$$

when heritability $(h^2) = 0.25$.

The accuracy of evaluation when m paternal half-sib progeny are used is

$$m/[m+(4-h^2)/h^2]^{1/2} = [m/(m+15)]^{1/2}$$

for $h^2 = 0.25$.

The accuracy with small numbers of "selfs," as compared to paternal half-sib progeny, is large and would need to be considered when balanced against costs of regular test matings as compared to selfing. Table V lists numerical values for accuracy of bull evaluations for selfing and for paternal sib progeny when heritability is 0.25.

Intensity of selection may not be affected with use of "selfs" unless more bulls were tested, which would be possible since fewer females in the population would be required to test each bull. The fraction of the population bred to young bulls could be reduced. Optimization of these factors would be required.

Mating of a bull to clones to obtain daughters would be equivalent to an evaluation of full-sib progeny except that the daughters could be about the same age. Contemporary full-sib daughters could also be obtained by embryo transfer. Table V shows, however, that full-sib progeny are not efficient for sire evaluation. In fact, if there are dominance effects or if an environmental correlation among full-sibs exists, then the accuracy of evaluation would be even less than shown in Table V.

Mating a bull back to clones of his mother or to her by superovulation and embryo transfer does not seem to be a very effective way of evaluating a bull, especially if the inbreeding difficulties are considered. If the mother is noninbred, the daughters of the bull are related by three-fourths to him and to each other but are also 25% inbred. If the original dam had been completely

TABLE V

Accuracy of Evaluating Sires from "Selfs," Paternal Half-Sib Progeny, Full-Sib Progeny, and When Mated to Clones of the Mother[a,b]

		Sib progeny		Son of clone mated to clones[b]	
Number (m)	"Selfs" $\left(\dfrac{m}{m+3.5}\right)^{1/2}$	Paternal $\left(\dfrac{m}{m+15}\right)^{1/2}$	Full $\left(\dfrac{m}{2m+14}\right)^{1/2}$	Noninbred $\left(\dfrac{9m}{12m+56}\right)^{1/2}$	Inbred clone $\left(\dfrac{4m}{5m+17}\right)^{1/2}$
1	0.47	0.25	0.25	0.36	0.43
5	0.77	0.50	0.46	0.62	0.69
10	0.86	0.63	0.54	0.72	0.77
20	0.92	0.76	0.61	0.78	0.83
50	0.97	0.88	0.66	0.83	0.87
100	0.98	0.93	0.68	0.85	0.88
∞	1.00	1.00	0.71	0.87	0.89
Inbreeding of test progeny	0.50	0	0	0.25	0.50

[a] Heritability, 0.25.

[b] Also equivalent to mating to superovulated mother followed by embryo transfer.

inbred by some form of nuclear manipulation, then the daughters of the bull would be related by 100% to him (similar to "selfing"), by 125% to each other (additive or numerator relationship) but would be 50% inbred. As shown in Table V, neither of these methods would be as effective as "selfing" to evaluate a bull.

Although selfing may have usefulness in biological research and limited use in testing for lethals, there seems to be no usefulness in selection for production.

VIII. CONCLUSIONS

A. Artificial Insemination

1. Much genetic gain for milk yield has been achieved through extensive use of bulls determined to be best from progeny evaluation.

2. The actual gain, however, is much less than theoretically possible. Selection of dams of sons has not been as effective as theory would indicate, probably due to biases of some type. Methods of obtaining unbiased records of potential bull dams are needed.

3. Selection for sires of sons and sires of cows has not been nearly as intense as recommended according to theory.

4. Emphasis on traits other than milk yield has further decreased the effective selection intensity for milk yield.

5. Proper application of AI could increase genetic progress for milk yield to a rate 4 times what has occurred.

B. Sexing Semen

1. Individual dairymen might profit by increased opportunities for selection of dams of cows. A positive net return over investment would depend on a relatively low charge for sexing semen ($10-$20 per breeding unit).

2. Dams of sons might be selected more intensely, but as stated above, evaluation of dams has not been a dependable indicator of the son's genetic value.

3. Numbers of sires of sons and of sires of cows required with AI would be relatively unchanged with some possibility that additional numbers would be needed.

C. Embryo Transfer

1. Considerable potential for genetic improvement exists for individual dairymen from selection of heifer replacements from a small fraction of the top cows. An economic loss from investment in embryo transfers is likely, however, due to apparently high minimum costs of transfer. The marginal cost based on genetic improvement is $50 to $90 per conception.

2. Increased selection intensity for dams of sons is possible, but the difficulty with dams of sons evaluations must still be overcome.

D. Cloning

1. An initial large gain in improvement could be made by selection of cows with high phenotypic records. Adequacy of quantitative genetic theory needs to be determined for such extreme selection intensity.

2. Methods of obtaining further gain need to be developed.

3. Duplication of random gametes before cloning would quickly expose lethal recessive genes. If any such inbred animals would survive, recessive lethals could be easily eliminated.

E. Selfing

1. Selfs obtained from combining random X- bearing gametes of a bull would be 50% inbred and would aid in eliminating recessive lethals.

2. The high relationships of the progeny to the bull would lead to high accuracy of genetic evaluation with relatively few progeny.

3. Nonlethal inbreeding effects, however, may create problems of evaluation for which solutions are not now available.

F. General

More precise systems analyses, which would combine various ranges of biological limits and costs, are needed to determine optimum breeding systems with sexed semen, embryo transfer, cloning, and selfing. Even if added net profit should exceed costs by only the equivalent of 1 lb of milk per cow and this gain were cumulative each year, in only 3 years this extra return would be over $1,000,000 when applied to the dairy population of the United States. Such a study should require little more than 1 or 2 years time of a postdoctoral person in systems analysis or animal breeding and a few thousand dollars (5 to 10) for computing and other costs. Such a study would require a relatively modest investment, bracket the economic potential of these reproductive systems, and be useful in deciding how many research dollars should be devoted to the various reproductive alternatives.

IX. APPENDIX: EQUATIONS FOR CALCULATING ACCURACY OF EVALUATION ACCORDING TO SELECTION INDEX THEORY

Assumptions

1. Only additive genetic effects contribute to likeness between records of relatives, that is, $Cov(X_iX_j) = a_{ij}h^2\sigma^2$, where X_i and X_j are records of relatives

i and j, a_{ij} is the additive or numerator relationship, h^2 is heritability in the narrow sense, and σ^2 is the phenotypic variance.

2. There is no environmental covariance between records of relatives.

3. Repeatability (r) is the correlation between records on the same animal.

1. Cow Evaluation

1. n records on the cow

$$\frac{1+(n-1)r}{n}\, b = h^2; \quad r_{G\hat{G}} = \sqrt{b} = \sqrt{\frac{nh^2}{1+(n-1)r}}$$

2. Single records on m full-sib progeny of the cow

$$\frac{1+(m-1)h^2/2}{m}\, b = h^2/2; \quad r_{G\hat{G}} = \sqrt{b/2} = \tfrac{1}{2}\sqrt{\frac{mh^2}{1+(m-1)h^2/2}}$$

3. n records on the cow; single records on m full sib progeny of the cow

$$\frac{1+(n-1)r}{n}\,(b_1) + (h^2/2)b_2 = h^2$$

$$(h^2/2)\,(b_1) + \frac{1+(m-1)h^2/2}{m}\, b_2 = h^2/2; \quad r_{G\hat{G}} = \sqrt{b_1 + b_2/2}$$

4. Single records on m clones of the cow

$$\frac{1+(m-1)h^2}{m}\, b = h^2; \quad r_{G\hat{G}} = \sqrt{b} = \sqrt{\frac{mh^2}{1+(m-1)h^2}}$$

5. n records on the cow; single records on m clones of the cow

$$\frac{1+(n-1)r}{n}\,(b_1) + h^2\,(b_2) = h^2$$

$$h^2(b_1) + \frac{1+(m-1)h^2}{m}\,(b_2) = h^2; \quad r_{G\hat{G}} = \sqrt{b_1 + b_2}$$

6. Single records on m paternal half-sibs of the cow

$$\frac{1+(m-1)h^2/4}{m}\, b = h^2/4; \quad r_{G\hat{G}} = \sqrt{b/4}$$

7. n records on the cow; single records on m paternal half-sibs of the cow

$$\frac{1+(n-1)r}{n}\,(b_1) + (h^2/4)\,(b_2) = h^2$$

$$(h^2/4)\,(b_1) + \frac{1+(m-1)h^2/4}{m}\,(b_2) = h^2/4; \quad r_{G\hat{G}} = \sqrt{b_1 + b_2/4}$$

2. Sire Evaluation

1. Single records on m half-sib progeny

$$\frac{1+(m-1)h^2/4}{m}\,(b) = h^2/2; \quad r_{G\hat{G}} = \sqrt{b/2}\ \sqrt{\frac{m}{m+\dfrac{4-h^2}{h^2}}}$$

2. Single records on m full-sib progeny

$$\frac{1+(m-1)h^2/2}{m} \quad (b) = h^2/2; \quad r_{G\hat{G}} = \sqrt{b/2} = (\tfrac{1}{2})\sqrt{\frac{mh^2}{1+(m-1)h^2/2}}$$

3. Single records on m "selfs" (sire furnishes both gametes); inbreeding of "selfs" = F = $\tfrac{1}{2}$; additive relationships of selfs = 1; additive relationships of "selfs" to sire = 1

$$\frac{1+Fh^2 + (m-1)h^2}{m} \quad (b) = h^2; \quad r_{G\hat{G}} = \sqrt{b} = \sqrt{\frac{m}{m+\left(\frac{2-h^2}{2h^2}\right)}}$$

4. Single records on m daughters of a sire mated to clones of his mother; inbreeding of daughters = F = $\tfrac{1}{4}$; additive relationships of daughters = $\tfrac{3}{4}$; additive relationships of daughters to sire = $\tfrac{3}{4}$

$$\frac{1+(m-1)\,(3/4)h^2}{m} \quad (b) = (3/4)h^2; \quad r_{G\hat{G}} = \sqrt{(3/4)b}$$

5. Single records on m daughters of a sire mated to clones of his completely inbred mother; inbreeding of mother = 1; inbreeding of daughters = F = $\tfrac{1}{2}$; additive relationships of daughters = 5/4; additive relationships of daughters to sire = 1

$$\frac{1 + Fh^2 + (m-1)(5/4)h^2}{m} \quad (b) = h^2; \quad r_{G\hat{G}} = \sqrt{b}$$

ACKNOWLEDGMENTS

The advice and suggestions of Drs. George Seidel and Robert Foote are greatly appreciated.

REFERENCES

Bichard, M. (1971). *Anim. Prod.* **13**, 401–411.

Bichard, M., Pease, A. H. R., Swales, P. H., and Özkütük, K. (1973). *Anim. Prod.* **17**, 215–227.

Brascamp, E. W. (1973). *Z. Tierz. Zuchtungsbiol.* **90**, 126–140.

Butcher, K. R., and Legates, J. E. (1976). *J. Dairy Sci.* **59**, 137–152.

Cunningham, E. P. (1976). *In* "Egg Transfer in Cattle" (L. E. A. Rowson, ed.), pp. 345–353. Luxembourg.

Dickerson, G. E., and Hazel, L. N. (1944). *J. Agric. Res.* **69**, 459–476.

Everett, R. W. (1975). *J. Dairy Sci.* **58**, 1717–1722.

Foote, R. H., and Miller, P. (1971). *In* "Sex Ratio at Birth-Prospects for Control" (C. A. Kiddy and H. D. Hafs, eds.), pp. 1–9. ASAS Symposium.

Hansen, M. (1976). *In* "Egg Transfer in Cattle" (L. E. A. Rowson, ed.), pp. 369–376. Luxembourg.

Henderson, C. R. (1954). *Proc. 7th Ann. Conf. Natl. Assoc. Artif. Breeders*, pp. 93–103.

Henderson, C. R. (1964). *J. Dairy Sci.* **47**, 439–441.

242   L. Dale Van Vleck

Henderson, C. R. (1973). *Proc. Anim. Breeding Genet. Symp. in Honor of Dr. Jay L. Lush, July 29, 1972, Blacksburg, Va.*, p. 10.

Henderson, C. R. (1975). *Biometrics* **31**, 423–447.

Henderson, C. R., and Dunbar, R. S. (1952). *Farm Res.* **18**, 3.

Hill, W. G. (1971). *Anim. Prod.* **13**, 37–50.

Hill, W. G. (1974). *Anim. Prod.* **18**, 117–139.

Hill, W. G., and Land, R. B. (1976). *In* "Egg Transfer in Cattle" (L. E. A. Rowson, ed.), pp. 355–363. Luxembourg.

Hinks, C. J. M. (1971). *Anim. Prod.* **13**, 209–218.

Hunt, M. S., Burnside, E. B., Freeman, M. G., and Wilton, J. W. (1974). *J. Dairy Sci.* **57**, 251–257.

Kräusslich, H. (1976). *In* "Egg Transfer in Cattle" (L. E. A. Rowson, ed.), pp. 333–342. Luxembourg.

Land, R. B., and Hill, W. G. (1975). *Anim. Prod.* **21**, 1–12.

Meadows, C. E. (1974). *J. Dairy Sci.* **57**, 626 (Abstr.).

Petersen, P. H., and Hansen, M. (1977). *Livestock Prod. Sci.* **4**, 305–312.

Rendel, J. M., and Robertson, A. (1950). *J. Genet.* **50**, 1–8.

Robertson, A., and Rendel, J. M. (1950). *J. Genet.* **50**, 21–31.

Searle, S. R. (1961). *J. Dairy Sci.* **44**, 1103–1112.

Skjervold, H. (1963). *Acta Agric. Scand.* **13**, 131–140.

Skjervold, H. (1966). *Proc. 9th Int. Congr. Anim. Prod., Edinburgh,* p.250.

Skjervold, H. (1974). *Symp. Egg Transplantation:* Publ. 87. SHS, Hallsta, Sweden.

Skjervold, H., and Langholz, H. J. (1964). *Z. Tierz. Zuchtungsbiol.* **80**, 25–40.

Specht, L. W., and McGilliard, L. D. (1960). *J. Dairy Sci.* **43**, 63–75.

Vaccaro, R. (1974). Ph.D. Thesis, Cornell University, Ithaca, New York.

Van Vleck, L. D. (1964). *J. Dairy Sci.* **47**, 441–446.

Van Vleck, L. D. (1969). *J. Dairy Sci.* **52**, 768–774.

Van Vleck, L. D. (1975). *Hoard's Dairyman*, Aug. 25, p. 950.

Van Vleck, L. D. (1976). *In* "Genetics Research 1975–1976," Report to Eastern Artificial Insemination Coop., Inc., p. 73. Dept. of Anim. Sci., Cornell Univ., Ithaca, New York.

Van Vleck, L. D. (1977). *In* "Proc. International Conference on Quantitative Genetics," p. 543. Iowa State Univ. Press, Ames.

Van Vleck, L. D., and Carter, H. W. (1972). *J. Dairy Sci.* **55**, 214–217.

Van Vleck, L. D., and Everett, R. W. (1976). *J. Dairy Sci.* **59**, 1802–1807.

Vinson, W. E. (1975). *J. Dairy Sci.* **58**, 1071–1077.

Wilmut, I., and Hume, A. (1978). *Vet. Rec.* **103**, 107–110.

Wilton, J. W., Evans, D. A., and Van Vleck, L. D. (1968). *Biometrics* **24**, 937–949.

13

Economic Benefits of Reproductive Management, Synchronization of Estrus, and Artificial Insemination in Beef Cattle and Sheep

E. KEITH INSKEEP AND JOHN B. PETERS

NEW TECHNOLOGIES IN ANIMAL BREEDING
Copyright © 1981 by Academic Press, Inc.
All rights of reproduction in any form reserved.
ISBN 0-12-123450-9

I. INTRODUCTION

The estimated 47.8 million cows and 8.2 million breeding ewes in the United States are of particular value in an energy conscious economy because they convert forages and by-products that are of little direct use to man into high quality food, especially protein. Bellows *et al.* (1979) have pointed out that beef cattle rank among the top five producers of gross income among agricultural commodities in 47 states, and they are first in cash receipts in 21 states. Yet the cow-calf operation has probably averaged a net loss in most states for most of the last 30 years (Barr, 1966; Barr *et al.*, 1967). Barr *et al.* (1966) found that sheep were more profitable in West Virginia than beef cattle, yet sheep numbers have declined in that state, as well as nationally, from 1940 until 1977. The number of ewe lambs in the United States under 1 year of age on January 1, 1979, was 12% higher than a year earlier. With 1980 figures revealing a 2% increase overall in ewe numbers, it would appear that the decline has halted.

Productivity is far from optimum in both the beef cattle and the sheep industries. The processes of selection and of crossing specific strains, which have been so effective in poultry, dairy cattle, and hogs, have only begun to be applied in an effective manner in beef cattle and have been virtually ignored in sheep.

II. CURRENT STATUS OF THE BEEF INDUSTRY AND THE USE OF ARTIFICIAL INSEMINATION

A. Calf Crop—Current Status

Net calf crop (defined as marketable calves per cow exposed for breeding) has been estimated at 65–81% in the United States. Examining data for 12,827 cow years at Miles City, Montana, Bellows *et al.* (1979) reported a 71% calf crop, with 17.4 percentage points of the loss (60%) attributable to females not pregnant at the end of a 45–60 day breeding season using natural service. Other losses were perinatal calf deaths: 6.4 percentage points (22%), calf deaths from birth to weaning: 2.9 percentage points (10%), and fetal deaths during gestation: 2.3 percentage points (8%). In West Virginia University's Allegheny Highlands Project (AHP) calf crop, per cow on the farms (January 1) averaged 87.6 and 84.2% for two groups of approximately 28 farms each over 5–7 and 3–4 years, respectively (Baker *et al.*, 1978, 1979). In that study, 6 and 7%, respectively, of calves born died before weaning.

B. Calf Crop—Gains to Be Made with Current Technology

Gerrits *et al.* (1979) have estimated that a 5% increase in the national calf crop would yield a total savings of $588 million in producing the current total supply of United States-grown beef. Techniques now available can produce an increase of this magnitude when integrated into an adequate management program. In the AHP, performance testing, crossbreeding, artificial insemination (AI), and pregnancy testing have been promoted as part of a package of technological assistance to the hill land farmer. Adoption of these practices and construction of suitable handling facilities to carry them out has varied from farm to farm, but by 1978, calving percentage for 2668 cows present on January 1 had reached 94.5%. Thus the percentage of nonpregnant cows wintered was only 5.5% compared to the 17.4% reported by Bellows *et al.* (1979). Assuming a wintering cost of $150 per cow, the savings from such an improvement on a national scale would total $853 million. The AHP staff observed that other uses of the records and handling facilities, which are needed to put these genetic and reproductive techniques into practice, led to an overall improvement in management. Unfortunately, this has not yet led to a decrease in either perinatal or postnatal calf mortality. On those farms using pregnancy testing and culling of open cows in the fall, the proportion of open cows wintered has been reduced to 1.5%, a figure comparable to the estimate of fetal deaths (2.3%) by Bellows *et al.* (1979). Full adoption of pregnancy testing could potentially save a further $286 million on a national basis.

The calf crop could be increased and calves could be born earlier in the breeding season if replacement heifers were selected on the basis of the conception rate to first service of their dams. Conception rate to first insemination with semen from a high fertility bull has a heritability of 8–10% (Inskeep *et al.*, 1961; Collins *et al.*, 1962), but this trait has been ignored in selection programs. In fact, adjusted 205-day weight is recommended by most animal breeders as the single trait for selection. Participants in current performance testing programs are often selecting against both short postpartum intervals to estrus and conception rates by saving later-born calves, which were weighed for test purposes at less than 180 days of age. An excellent discussion of overall reproductive management needed to maximize calf crop has been presented by Bellows (1980).

C. Use of Artificial Insemination in Beef Cattle

Currently, less than 5% of first inseminations in beef cattle in the United States are by AI. Hafs (1979) cites a figure of 3.4% for 1971. One major limitation to use of AI is the low average conception rate to first service, which averages around 50% nationally. Beverly (1979) has reported that most beef producers in Texas expect a 60–70% conception rate to first service if they are to use AI.

D. Gains in Calf Growth to Be Expected with Artificial Insemination

The standardized measure of weaning weight in beef cattle is the weight at 205 days adjusted for sex of calf and age of dam. The average 205-day weight of calves from performance-tested herds in West Virginia was 467 lb in 1977, and market weights in feeder calf sales over the last 5 years averaged 464 lb for steers and 430 lb for heifers. In the last 2 years, market weights have increased due to the acceptance of crossbreeding; the 1979 values were 490 and 450 lb for steers and heifers, respectively. The number of beef bulls being performance-tested for weight gain to 205 days and in the feedlot has increased dramatically in recent years. Gain to 205 days has a heritability of 30% and increases in calf weaning weight of 50 lb with straight breeding and 100 lb with crossbreeding are possible in a single generation. In the AHP, emphasis has been placed upon selection of superior gaining bulls for use as herd sires in either natural mating or AI programs, along with improved nutrition, health, and over-all management. Increases in calf weights at 205 days have averaged 10 lb per year of participation in the project. This increase has been associated with an increase of 8 percentage points per year in the proportion of crossbred calves raised. Baker *et al*. (1978) have estimated that extending the increase achieved by AHP cooperators to all calves in the state would add $3,600,000 per year to the income of West Virginia beef producers when calf prices averaged $50 per hundredweight. Rapid adoption of AI could bring about this kind of increase in income within as little as 40–48 months.

III. CURRENT TECHNOLOGY OF SYNCHRONIZATION OF
ESTRUS IN BEEF CATTLE

A number of authors have reviewed the methods for synchronization of estrus in beef cattle, most recently Hansel and Beal (1979), Lauderdale (1979), and Moody (1979). The use of progestogens was reviewed in detail at a conference at Lincoln, Nebraska, in 1964 (USDA Miscellaneous Publication 1005), and recent developments have been discussed by Wiltbank and Mares (1977). Particular attention is now directed to prostaglandin (PG) $F_2\alpha$, the luteolytic hormone that has recently been approved for use in dairy heifers and beef cattle. Using a treatment regimen of (1) rectal palpation of the ovaries, (2) treatment of cows that have corpora lutea in the ovaries with 25 mg $PGF_2\alpha$, and (3) injection of the

TABLE I

Proportion of a Lactating Beef Herd Pregnant after Synchronization and AI

Animals in estrus or with a palpable CL (%)	Treatment	Animals in synchronized estrus (%)	Conception rate (%)	Pregnant animals in herd (%)
56.6	$PGF_2\alpha$	74[a]	46[a]	19.3
	$PGF_2\alpha$ and EB	92[a]	48[a]	25.0
	SMB—cycling	88[b]	58[c]	28.9
	SMB—noncycling	60[d]	45[d]	11.7
			Total	40.6

[a] Based on data of Welch et al. (1975) and Woloshuk (1977).

[b] Based on data from 29 cows in West Virginia.

[c] Estimated as the median of values of 45–70% reported by Wiltbank and Mares (1977).

[d] Based on 84 cows in a West Virginia study.

treated cows with 400 μg estradiol benzoate 40–48 hours later (a treatment that is not on the market), it has been possible to obtain close synchrony of estrus in 92% of the treated animals (Table I) and conception rates averaging 50% to either breeding 12 hours after estrus or 80 hours after $PGF_2\alpha$, thus a net pregnancy rate of 46% of the cycling cows in the herd (Inskeep et al., 1980). Conception rate has varied with service sire, technician, and farm. Handling facilities and cow condition were major contributors to differences among farms.

A major limitation to use of this technique is the fact that not all cows are reproductively active at the onset of the breeding season. In herds studied over a 4-year period (Inskeep and Lishman, 1979), the proportion of cows either in heat or ready to synchronize with $PGF_2\alpha$ on the first day of the breeding season averaged only 56%. The other technique being readied for market, utilizing ear implants of progestogen for 9 days, is of some value in those cows and heifers that have not yet begun ovulatory cycles. Ideally, synchronization rates of 60% and conception rates of 40% might be obtained in the noncycling animals (Wiltbank and Mares, 1977). With temporary removal of calves coincident with removal of implants, these figures might be improved to 85 and 50%, respectively (for a comparison of the potential of $PGF_2\alpha$ and progesterone treatments see Table I). Nevertheless, there remains a critical need to understand the mechanisms that are involved in the attainment of early puberty and of short postpartum intervals to estrus and ovulation so that the proportion of noncycling animals may be reduced. It is known that breed, age, and nutritional status make significant contributions, but the mechanisms whereby these effects are asserted still require study (Inskeep and Lishman, 1979).

IV. ECONOMIC VALUE OF ARTIFICIAL INSEMINATION AT SYNCHRONIZED ESTRUS IN BEEF CATTLE

Synchronization of estrus is a tool to reduce the labor costs involved in the utilization of AI. The major deterrent to use of AI in beef cattle has been the need to detect estrus over long periods of time if a significant proportion of the herd was to become pregnant by AI. Peters *et al.* (1979) have attempted to determine whether artificial insemination is economically viable in the small herds, which make up most of the beef industry in the United States, when synchronization of estrus is used, and each cow is inseminated a second time if she does not conceive at first estrus, but no clean-up bulls are used. The system of breeding entirely by AI was compared to use of average or superior performance-tested bulls available for purchase through annual sales. They found that AI was very profitable in herds of 10 cows, marginally profitable in 20-cow herds (depending upon conception rate to first service), but probably not competitive with superior performance-tested bulls in herds of 30 cows, assuming that a single bull was sufficient for 30 cows in natural service. That study was done with 1975 calf prices.

A recalculation using 1979 calf prices recently was made by the methods used by Peters *et al.* (1979). In that case, the superior performance-tested bull and AI were nearly equal in 10-cow herds, a change from 1975 when for that herd size, average bulls had been second to AI. The superior bull appeared to be more economical than AI in 20- and 30-cow herds with both straight breeding and crossbreeding, using 1979 calf prices. Two factors that contribute to changes in the relative value of AI were noted. One was technician cost per cow, which is increasing and for which volume discounts have not yet been offered in all areas. The second was the relationship of the price of corn to that of cattle. When corn is cheap, prices per pound for lightweight calves increase more relative to those for heavy calves, so that there is less economic incentive to support genetic improvement in those years.

A weakness in the comparisons made by Peters *et al.* (1979) is that no added value was given to replacement heifers generated by superior bulls in natural service or AI. Estimates for this have ranged from only $15 (Singleton and Petritz, 1976) to as high as $150–175 (Wallace, 1979). Obviously all the costs and returns vary with farm and with the cattle cycle. So one can say only that the smaller the herd, the more cows cycling, the higher the conception rate, and the better the bull used, the more valuable AI becomes. None of the specific figures would be applicable to purebred herds, where much larger benefits can be expected from AI. In an earlier study (Stevens and Mohr, 1969), the estimated increased value per calf when AI was used was $30.02 on purebred ranches compared to $3.31 on commercial ranches in Wyoming.

Donaldson (1980) has done an economic study comparing several management schemes using $PGF_2\alpha$ to AI for two services (42 days) without benefit of synchronization. He estimated cost per calf for each system. It was $26 per calf for control cows bred 12 hours after observed estrus during 21 or 42 days and for cows bred 12 hours after detected estrus for 9 or 10 days and treated with 12.5 mg $PGF_2\alpha$ on day 5 if not in estrus by that time. The interval from calving to breeding was 14 days shorter for the synchronized cows. Doubling the dose of $PGF_2\alpha$ to the 25 mg commonly recommended in the United States increased costs to $41 per calf for first service only or $30 if two service periods were used. Use of two doses of $PGF_2\alpha$ 12 days apart, increased costs dramatically to $34 and $45 per calf with 12.5 mg and 25 mg $PGF_2\alpha$, respectively. Moving to timed insemination at first service with the two dose scheme further increased costs to $75 and $100 per calf for first services or $55 and $68 over two services for 12.5 and 25 mg $PGF_2\alpha$, respectively. A scheme with two timed inseminations at first service cost $97 and $160 per calf to first service or $70 and $83 over two services for the low and high doses of $PGF_2\alpha$, respectively.

One major efficiency factor was that the proportion of cows assigned to the AI program that were inseminated was 72% in normal AI and 74% in the 10-day management system with $PGF_2\alpha$. The latter figure is exactly equal to that obtained in the West Virginia studies with $PGF_2\alpha$ only (Table I). Donaldson did not compare a system utilizing estradiol benzoate with $PGF_2\alpha$, but if it increased the proportion of cows inseminated as in the West Virginia studies and some Australian work (Nancarrow and Radford, 1975), then it should reduce calf cost below control values.

Experience in the AHP and in various field trails has revealed that most producers, particularly those with larger herds, will use AI only at the first synchronized estrus at the beginning of the breeding season, then follow-up with bulls in natural service. If the proportion of cows cycling at the beginning of the breeding season is maximized, this practice combines maximum return to labor with minimum labor input and thus maximizes the economic value of AI. Our general observation has been that, in herds using synchronization of estrus, one or two services to AI, and culling of nonpregnant cows after a 60-day breeding season, the proportion of cows cycling at the beginning of the breeding season has increased over a 3–4 year period by as much as 20–30 percentage points. In the herd at West Virginia Reymann Memorial Farms, estrous synchronization and AI have been in use for 12 consecutive years, the breeding season is restricted to about 50 days, and no clean-up bulls have been used in the last 4 years. In 1980, 83 of the 99 cows were inseminated during the first 9 days of the breeding season, and 45 conceived to this service. Pride in the quality and weight of the calves obtained has increased the desire to continue using AI on most farms.

Hafs (1979) has predicted that the availability of $PGF_2\alpha$ could increase the number of beef calves born from superior bulls 10-fold and that perhaps 20% of the United States beef cow herd could receive at least one insemination artificially by 1990. If this led to a 50-lb increase in weight for 10% of the calves born, it should be worth \$114–122 million each year, assuming 80 or 85% net calf crop and \$60 per hundredweight. The spread effect of saving replacement heifers and commercial herd bulls from these matings could increase these figures by as much as 50%. This will happen only if veterinarians, AI technicians, and farmers learn the proper use and limitations of $PGF_2\alpha$, especially since it is not effective unless cows are showing regular cycles. For anovulatory cows, cyclic activity can be initiated by 9-day progestogen treatment. Further study is needed to determine whether this treatment should be used alone or followed in some way by $PGF_2\alpha$ (Wiltbank and Mares, 1977; Hansel and Beal, 1979).

The decision by a producer to use or not to use AI and synchronization of estrus in any given year will depend upon the relationships among several economic factors that have been discussed and upon his understanding of how these tools fit into his overall management program, including reproductive management (Bellows, 1980). Along with heat checking and low conception rates, lack of suitable handling facilities has limited the use of AI. The impetus to construct these facilities on AHP farms was the direct demonstration of the value of selection for 205-day weight and of use of AI, pregnancy testing, and other management techniques which these facilities made possible. This requires a one-to-one approach rather than the mass media techniques currently in vogue in many extension programs. Spread effect will be achieved only after successful examples exist in local farming communities.

V. CURRENT STATUS OF THE SHEEP INDUSTRY AND OF SYNCHRONIZED ESTRUS AND ARTIFICIAL INSEMINATION FOR SHEEP

Currently, the net lamb crop in the United States is estimated at 94–97%. In West Virginia, the state average is estimated at 112–115%, which is perhaps relatively typical for the farm flock states. Most lambs in the farm flock states are marketed at 90–115 lb; the West Virginia average marketable lamb per ewe is currently around 96 lb. A comparable figure for the entire United States is not available; there are tremendous annual, seasonal, and geographical differences in the proportions of lambs that go directly to slaughter or into feedlots. In areas such as West Virginia and Virginia, slaughter-weight lambs can be produced on milk and forage alone. Hulet (1977), at a symposium evaluating management of reproduction in sheep and goats, pointed out that "research has far outstripped application" of knowledge in the sheep industry.

A. Gains to Be Made in Lamb Crop and Growth Rate

The heritabilities of the two traits of most immediate importance to the United States sheep industry—weaning weight and number of lambs born—are 25 and

15%, respectively. Turner (1968) has shown that selection for number of lambs born can be very effective and has recommended that selection in sheep be based upon independent culling levels rather than a selection index. Bradford (1977) has reviewed how various reproductive traits can be subjected to selection and management and where further research is needed. The staff of the AHP has recommended selection of replacement ewes from among the fast-growing ewe lambs born as twins and the use of flushing to increase ovulation rates. Lambing percentage has increased by 1.8 percentage points and market weight of lambs by 1 lb per year of cooperation in the project. Thus by 1977, cooperators in the first phase of the project averaged 134% marketable lamb crop and 130 lb of marketable lamb per ewe. Baker *et al.* (1978) estimated that a similar improvement statewide would generate an additional $1,661,920 of income. At current lamb prices ($60 per hundredweight) increasing the national lamb crop by 20 lb per ewe would generate an additional $98.4 million of product from the current supply of ewes and enable a reduction of imports of lamb and a faster repopulation of breeding ewes.

B. Value of Synchronized Estrus

Synchronization of estrus in ewes can be done with any number of progestogens and is being utilized extensively in many countries (Gordon, 1977; Ainsworth *et al.*, 1977; Thimonier and Cognie, 1977), but there are not any products for this purpose currently marketed in the United States. Intravaginal sponges containing progestogen were available for some years and have been used in about 2000 ewes per year in West Virginia. In the ewe, synchronization of estrus is a labor saving method for management of lambing. High conception rates are obtained with ewe:ram ratios of 12:1 or lower, especially with breeding at the second estrus after treatment, and ewes that do not conceive to first breeding remain synchronized for at least two cycles. Ewes conceiving on the same day lamb within a 7-day period, and therefore those conceiving within 3 days of each other lamb during a 10-day span. Thus the part-time shepherd can plan his vacation for the first round of lambing and lamb out his entire flock in less than 30 days, with two 10-day intervening periods during which no lambs are born. This system has enabled two West Virginia shepherds to expand their flocks to 500 ewes, a number which can provide a net income of $12–15,000 per year.

Work is currently being done with $PGF_2\alpha$, and it is likely to become the treatment of choice within a few years. In a recent study on five farms in West Virginia (P. E. Lewis and E. K. Inskeep, unpublished), 236 ewes were treated with two doses of 20 mg $PGF_2\alpha$ 12 days apart, and 223 ewes received intravaginal sponges containing 20 mg fluorogestone acetate. Pregnancy rates to first service were 64 and 62% and lambing rates averaged 1.75 and 1.65, respectively.

Management of reproduction in sheep can be aided by greater knowledge of the stimulatory effect of the ram. Near the onset of the breeding season, estrus can be hastened by introduction of rams. Lewis and Inskeep (1973) maximized pregnancy rate to first service by using intravaginal sponges for 12 days and teaser rams during the first synchronized estrus; breeding occurred at the second estrus. This was effective for either AI or natural mating.

C. Use of Artificial Insemination

The technology for AI in sheep has been reviewed by Inskeep *et al.* (1974), Colas and Courot (1977), and in detail by Salamon (1976). Probably fewer than 1000 ewes per year are inseminated artificially in the United States. Use of AI has been limited by (1) the relatively poor fertility obtained in most trials with frozen semen and (2) the failure to identify truly superior performance and progeny-tested rams whose use justifies the costs of AI. In the early years (1971–1974) of the AHP, performance-tested rams were purchased and used for AI at synchronized estrus in several flocks. Conception rates were satisfactory with fresh semen (Inskeep *et al.*, 1974), but 90- or 120-day weights of lambs from those rams were not consistently better than those of lambs from less expensive rams that producers could obtain locally. Crossbred ewe lambs from some of these matings were retained, but their value became apparent only as they produced higher lambing percentages as 3- and 4-year-olds and exhibited greater longevity than their native counterparts. By that time, farmers had lost interest in the use of AI. More rapid progress was achieved by crossbreeding and by selection of replacement ewes born as twins, coupled with management steps to control parasites and to increase forage production.

In our judgement, AI will not be used in sheep until adequate systems for performance and progeny testing for number of lambs born and raised are devised, until selection for growth rates is implemented more extensively, and until routine freezing and storage of ram semen can be done in a reliable manner. Currently, selection within the flocks for twinning and effective techniques of predator control will probably return the producer's investment more rapidly than any other new steps that might be taken in management. As ways to improve performance and progeny testing are developed and implemented, and reliable methods for overcoming anestrus become available, use of AI at synchronized estrus will become more valuable.

VI. CONCLUSIONS

There are numerous clear demonstrations that knowledge of reproductive physiology can be applied in management and selection of beef cattle and sheep to increase production per unit cost. That applications have lagged behind acquisition of knowledge is clear, and it appears that a greater one-to-one effort in extension education is needed. Methods to synchronize estrus are now a powerful tool in the implementation of AI programs in beef cattle, but they must be used judiciously within the limits of other management factors. Further work is needed in both species, particularly in understanding anestrus and delayed puberty.

ACKNOWLEDGMENTS

The authors acknowledge the assistance of Alfred L. Barr, Barton S. Baker, and Paul E. Lewis in reading and in discussing this manuscript and in allowing the use of their data. Support for studies at West Virginia University was provided by grants to the West Virginia Agricultural and Forestry Experiment Station from The Rockefeller Foundation and by Hatch Project 224.

REFERENCES

Ainsworth, L., Hackett, A. J., Heaney, D. P., Langford, G. A., and Peters, H. F. (1977). *Proc. Symp. Management Reprod. Sheep Goats., Madison, Wisc.* pp. 101–108.

Baker, B. S., Fausett, M. R., Woodson, F. E., Lewis, P. E., Colyer, D. K. and Inskeep, E. K. (1978). A Progress Report on the Allegheny Highlands Project: Agriculture, January–December 1977, West Virginia University, Morgantown, 72 pp.

Baker, B. S., Fausett, M. R., Lewis, P. E., and Inskeep, E. K. (1979). A Progress Report on the Allegheny Highlands Project: Agriculture, January–December 1978, West Virginia University, Morgantown, 72 pp.

Barr, A. L., (1966). *W.Va. Agric. Exp. Stn. Bull.* **527.**

Barr, A. L., Wamsley, B. W., Jr., and Templeton, M. C. (1966). *W.Va. Agric. Exp. Stn. Bull.* **495** (revised).

Barr, A. L., Toben, G. E., and Wilson, C. C., Jr. (1967). *W.Va. Agric. Exp. Stn. Bull.* **546.**

Bellows, R. A., Short, R. E., and Staigmiller, R. B. (1979). *Anim. Reprod. BARC Symp.* **3**, 3–18.

Bellows, R. A. (1980). *Angus J.* January, 1980, pp. 84–90.

Beverly, J. R. (1979). *In* "Proc. Lutalyse Symposium, Brook Lodge" (J. W. Lauderdale and J. H. Sokolowski, eds.), pp. 81–84. Upjohn, Kalamazoo.

Bradford, G. E. (1977). *Proc. Symp. Management Reprod. Sheep Goats. Madison, Wisc.* pp. 1–19. (Appendix).

Colas, G., and Courot, M. (1977). *Proc. Symp. Management Reprod. Sheep Goats, Madison, Wisc.* pp. 31–40.

Collins, W. E., Inskeep, E. K., Tyler, W. J., and Casida, L. E. (1962). *J. Dairy Sci.* **45**, 1234.

Donaldson, L. E. (1980). *Theriogenology.* **14**, 391–401.

Gerrits, R. J., Blosser, T. H., Purchase, H. G., Terrill, C. E., and Warwick, E. J. (1979). *Anim. Reprod. BARC Symp.* **3**, 413–421.

Gordon, I. (1977). *Proc. Symp. Management Reprod. Sheep Goats, Madison, Wisc.* pp. 15–30.

Hafs, H. D. (1979). *In* "Proc. Lutalyse Symposium, Brook Lodge." (J. W. Lauderdale and J. H. Sokolowski, eds.), pp. 9–14. Upjohn, Kalamazoo.

Hansel, W., and Beal, W. E. (1979). *Anim. Reprod. BARC Symposium* **3**, 91–110.

Hulet, C. V. (1977). *Proc. Symp. Management Reprod. Sheep Goats, Madison, Wisc.* pp. 119–133.

Inskeep, E. K., and Lishman, A. W. (1979). *Anim. Reprod. BARC Symp.* **3**, 277-289.

Inskeep, E. K., Tyler, W. J., and Casida, L. E. (1961). *J. Dairy Sci.* **44**, 1857.

Inskeep, E. K., Stevens, J. T., Peters, J. B., and Kauf, L. (1974). *W.Va. Agric. Exp. Stn. Bull.* **629**.

Inskeep, E. K., Dailey, R. A., James, R. E., Peters, J. B., Lewis, P. E., and Welch, J. A. (1980). *Proc. 9th Int. Congr. Anim. Reprod. Artif. Insem. Madrid,* **3**, 138.

Lauderdale, J. W. (1979). "Proc. Lutalyse Symposium, Brook Lodge" (J. W. Lauderdale and J. H. Sokolowski, eds.) pp. 17-32. Upjohn, Kalamazoo, Michigan.

Lewis, P. E., and Inskeep, E. K. (1973). *J. Anim. Sci.* **37**, 1195-1200.

Moody, E. L. (1979). *In* "Proc. Lutalyse Symposium, Brook Lodge" (J. W. Lauderdale and J. H. Sokolowski, eds.), pp. 33-41. Upjohn, Kalamazoo.

Nancarrow, C. D. and Radford, H. M. (1975). *J. Reprod. Fertil.* **43**, 404.

Peters, J. B., Welch, J. A., Barr, A. L., and Inskeep, E. K. (1979). *In* "Hill Lands" (J. Luchok, J. D. Cawthon, and M. J. Breslin, eds.), pp. 631-635.

Salamon, S. (1976). "Artificial Insemination of Sheep," 104 pp. Publicity Press, Ltd., Chippendale N.S.W.

Singleton, W. L. and Petritz, D. C. (1976). The Charolais Way, May, 1976. pp. 55-65.

Stevens, D. M. and Mohr, T. (1969). *Wyo. Agric. Exp. Stn. Bull.* **496**.

Thimonier, J., and Cognie, Y. (1977). *Proc. Symp. Management Reprod. Sheep Goats, Madison, Wisc.* pp. 109-118.

Turner, H. N. (1968). *Proc. Symp. Reprod. Sheep, Stillwater, Okla.* pp. 67-93.

Wallace, Roy. (1979). *In* "Proc. Lutalyse Symposium, Brook Lodge" (J. W. Lauderdale and J. H. Sokolowski, eds.), p. 97. Upjohn, Kalamazoo.

Welch, J. A., Hackett, A. J., Cunningham, C. J., Heishman, J. D., Ford, S. P., Nadaraja, R., Hansel, W., and Inskeep, E. K. (1975). *J. Anim. Sci.* **41**, 1686-1692.

Wiltbank, J. N., and Mares, S. E. (1977). *Proc. 11th Conf. Artif. Insem. Beef Cattle, Denver,* pp. 57-65.

Woloshuk, J. M. (1977). M. Agric. Problem Report—Application of $PGF_2\alpha$ and Estradiol Benzoate to Synchronize Estrus in Beef Cattle, West Virginia University, Morgantown.

V

Continuation and Implementation of Research

14

Continuation and Implementation of Research

BENJAMIN G. BRACKETT, GEORGE E. SEIDEL, JR., AND SARAH M. SEIDEL

I. BIOMEDICAL AND AGRICULTURAL RESEARCH

The state of the biological sciences has recently been described as a revolution, which places mankind on the threshold of some unusual transformations in health practices, agriculture, and industry (Fredrickson, 1980). Knowledge of living things is expanding geometrically. This reflects the commitment of serious public support for research in the natural sciences, especially during the last 30 years. Since 1950, approximately 75 billion dollars from private and public funds have been expended for research and development in health care in the United States, and the annual investment currently represents about 0.31% of the gross national product (Fredrickson, 1980). About 60% of total funds and an estimated 90% of the basic research have been supported by the Federal budget. Products of this research have included hormonal contraception, eradication of smallpox by

257

NEW TECHNOLOGIES IN ANIMAL BREEDING

vaccination, control of polio, rubella, and Rh disease by immunization, decreased mortality from cardiovascular disease, and elucidation of the molecular basis of many "inborn errors of metabolism." Regarding the latter, the list of human genetic disorders in which the defective gene product is known grew from 15 in 1960 to 1000 in 1980, and identification is now possible for the chromosomal location of more than 35 such mutant human genes. The entire human genome (more than 100,000 genes) may be mapped before the twenty-first century (Fredrickson, 1980).

Monetary inflation, decreased growth of industrial productivity, and critical shortages in energy represent threats to the size and vigor of scientific inquiries today by undermining the affluence of developed countries. According to Fredrickson (1980), if the biological revolution, now so well launched, is to be sustained through the 1980s and beyond, a sine qua non is increased attention to the government–university interaction in science. The National Institutes of Health, the layers of government above it, and members of the academic-scientific community must work together to maintain this effective partnership.

Possibly the private sector can compensate for declining governmental support of research and related training at universities. Industry requires a continuously renewed source of trained personnel and new information of many kinds. It may be more attractive to pay universities for these directly instead of indirectly through taxes. For example, trainees could make commitments to work in certain industries for specified periods in return for financial support. While such arrangements already exist, they are not nearly as common with private as with public sponsors (e.g., military support of Reserve Officer trainees, commitments after sabbatical leaves, agreements of Public Health Service trainees to work in teaching and research, and foreign student support from their governments).

A slowdown in growth of (labor) productivity has been a major concern in recent years (Council of Economic Advisors, 1979). Research leading to technological innovation has been recognized historically as the primary source of increases in productivity (Denison, 1962; Mansfield, 1972). Policies that have been suggested for accelerating innovation include government research grants, government laboratories, and tax credits (Mansfield, 1972). In reviewing economic benefits from public research in agriculture, Evanson et al. (1979) reached the following conclusions. Agricultural productivity continues to grow. Annual rates of return on research expenditure are on the order of 50%. This successful track record was attributed to the system of state agricultural experiment stations and substations along with close association of research oriented toward science with that oriented toward technology and farming. However, the system remains undervalued; the high rate of return shows that investment in public research in agriculture is too low. In a representative year (1975), total investment in research to improve livestock (cattle, swine, sheep, poultry, and dairy) was only 0.34% of the total gross income from this important class of

agricultural products. The undervaluation of agricultural research has been attributed to the spillover of research benefits to other regions (i.e., across state
lines to those who do not pay for the research) and to consumers (in such small
amounts that individual consumers fail to see the connection), reducing the
incentives for local (state) support (Evanson *et al.*, 1979).

Beattie (1980), in advocating a national agricultural policy, points out that
every successful civilization has been predicated upon a successful agriculture,
and he cautions that, if appropriate steps are not taken soon, food and agriculture
in the United States will become as serious a problem by the turn of the century
as inflation, unemployment, and energy are today. In 1940, 40% of the Federal
budget for research and development was devoted to agriculture, whereas the
comparable value today is only about 1.3%. Less than 3% of the United States
population is presently involved in production agriculture. The success of
American agriculture might be its own worst enemy since a loss of visibility has
accompanied increased efficiency, and this has resulted in a much smaller voice
in public policy and decision making. Clearly, agriculture has not solved the
problem of undervaluation of public research.

II. CONTINUATION OF RESEARCH IN ANIMAL REPRODUCTION

Without earlier advances in veterinary medicine and animal science (especially in areas of disease control, preventive medicine and herd health programs,
animal husbandry, and nutrition), the newer and more sophisticated technologies
could not be contemplated for practical usage. Many of the innovative
technologies in animal breeding have arisen from human health-related research
efforts made possible by research grants from the National Institutes of Health.
Basic biological information on male and female reproductive systems and pregnancy is necessary for improved approaches to important human problems including contraception (prevention of conception as an alternative to abortion),
male or female infertility, and environmental or genetic threats to gestational
development.

All of science is interrelated. Opportunities for research in animal reproduction
with greater freedom for innovation (in contrast to narrowly targeted research)
should pay off handsomely in the enhancement of domestic animal production
and, hence, in human nutrition, even though the original intent of such research
might have been to develop the basis for better human contraception. Similarly,
unfettered research in reproduction for application to animal agriculture could
lead to important advances in obstetrics and gynecology or urology.

Ethical constraints inherent in clinical management of human reproduction are
quite different for theriogenology. Acceptable limits in manipulations to control

animal reproduction, within accepted practices to insure humane management and treatment, are yet to be defined. The potential for good is so great that it would appear prudent for a society to avoid permitting ethical considerations that pertain only to human beings to hinder progress in innovative and practical developments in animal reproduction.

The convergence of basic studies in mammalian reproductive biology with livestock breeding is occurring for all classes of livestock, but especially for cattle. A more important future role in the agricultural livestock economy of the world may be assumed by small ruminants that share with the cow the ability to thrive on roughage with minimal requirements for energy derived from fossil fuels or competition with mankind for foodstuffs. An even greater advantage is offered by aquaculture, since fish require less energy for protein synthesis than land animals, and they can use resources unsuitable for other agriculture. The efficiency of food conversion of poultry insures a niche in the future of agriculture for this important commodity.

III. IMPLEMENTATION OF NEW TECHNOLOGIES

Many of the recent technological developments in reproductive biology will be applied in animal breeding before the turn of the century. Use of artificial insemination in animal breeding has proven to have the greatest impact of all the reproductive technologies. Development and application of artificial insemination in many species will continue to play a prominent role. Already apparent is greater application in cattle coincident with effective means of estrus synchronization. Embryo transfer will become more important with further development of ancillary technologies including, but not limited to, sex selection, *in vitro* fertilization, cryopreservation of embryos, and twinning in cattle. A continuation of the trends, suggested in preceding pages, should allow combined applications of complementing technologies in the twenty-first century in ways not possible to foresee today. It is likely that fewer animals will be required to produce equivalent quantities of animal products. This has tremendous potential for decreasing pressure on resources and the environment, and it also permits allocation of more resources for animals that are not used primarily for food and fiber but that greatly enhance the quality of life. An example is the rapidly increasing number of horses.

The "new biology," especially the uses of monoclonal antibodies and a variety of approaches to genetic engineering, will directly or indirectly affect all animals on earth in the next century. The rush of investment capital into recombinant DNA technology and the isolation or production of interferons, which are likely to be followed soon by commercialization of antibody production, has demonstrated in a dramatic way that industry stands ready to exploit developments of biological research. Impetus was provided by the recent Supreme Court decision that new forms of life are patentable. Only through unlimited freedom of communication has it been possible to transcend the anxiety evoked by recombinant DNA technology. New reproductive technologies once developed and applied in animals can clearly be expected to raise similar feelings of anxiety over possible implications of their use in *Homo sapiens*. Hopefully, the lessons learned through previous encounters with recombinant DNA and more recently with *in vitro* fertilization will, at least, encourage dialogue as a most important early step toward societal decision making.

Implementation of new technologies in animal breeding depends on interactions between many factors, beginning with a decision to strive for improvement, then proceeding with research and development to demonstrate feasibility, followed, in turn, by acceptance by breeders after long-term and short-term economic, genetic, and other considerations. A great deal of influence on this process of implementation is wielded by regulatory agencies, extension services, competitive commercial interests, and disinterested popular opinion. For example, agencies charged with decreasing health hazards involved in international movement of livestock may not appreciate the greatly reduced risks in transporting embryonic rather than adult animals, thereby defeating their own purpose by instituting inappropriate regulations discouraging trade in embryos. In the other direction, active extension programs may promote new technologies to the point that individuals who do not adopt them cannot compete successfully. Another complexity is that groups who speak in favor of animal rights may reverse trends toward intensive management, upon which many newer technologies depend, and thus delay the implementation of advances that would increase production. Commonly shared anthropomorphic ideas may actually be counter to the health and happiness of animals. On the other hand, more attention to the comfort of animals will likely result in more profitable production.

Increasing human populations with concomitant increasing needs for animals and animal products should insure continuing demands in the marketplace. Since many aspects of the chain culminating in implementation of new animal breeding technologies are directly influenced by societal decisions, farsighted planning, especially in research and development, must be a constant, ongoing process to maximize benefits to society.

REFERENCES

Beattie, J. M. (1980). *Health Affairs (Univ. of Pa.)* **16,** 13–15.

Council of Economic Advisors (1979). "Annual Report, 1979." Government Printing Office, Washington, D.C.

Dennison, E. F. (1962). "The Sources of Economic Growth in the United States and the Alternatives Before Us," Suppl. paper 13. Committee for Economic Development, New York.

Evanson, R. E., Waggoner, P. E., and Ruttan, V. W. (1979). *Science (Washington, D.C.)* **205,** 1101–1107.

Fredrickson, D. S. (1980). *N. Engl. J. Med.* **304,** 509–517.

Mansfield, E. (1972). *Science (Washington, D.C.)* **175,** 477–486.

Index